Least-Cost Energy
Solving the CO_2 Problem

Amory B. Lovins
L. Hunter Lovins
Florentin Krause
Wilfrid Bach

Brick House Publishing Company
Andover, Massachusetts

Brick House Publishing Co., Inc.
34 Essex Street
Andover, Massachusetts

Production Credits:
Editor: Jack Howell
Edited by Nancy Irwin
Copyedited by Joyce Thompson
Drawings by Terry LeBlanc
Designed and Produced by Mike Fender

Printed in the United States of America

Copyright © 1981 by Amory B. Lovins and L. Hunter Lovins
All rights reserved

No part of this book may be reproduced in any form without the written permission of the publisher.

Library of Congress Cataloging in Publication Data

Main entry under title:

Least-cost energy.

   Bibliography: p.
   1. Power resources. 2. Energy conservation. 3. Carbon dioxide—Environmental aspects. 4. Energy policy. I. Lovins, Amory B., 1947-
TJ163.2.L4   333.79   81-21584
ISBN 0-931790-36-0   AACR2

Contents

List of Figures	iv
List of Tables	v
Foreword	vii
STEPHEN H. SCHNEIDER, WALTER ORR ROBERTS	
Preface	xiii
Executive Summary	xvii

1 INTRODUCTION   1

    1.1. Energy/climate risks: a future problem whose time has passed? 1

    1.2. How much time can we buy? 8

2 HOW MUCH ENERGY WILL WE NEED?   13

    2.1. Defects of conventional forecasts 15

    2.2. Need for improved forecasts 22

    2.3. Case study for the Federal Republic of Germany 24

       2.3.1. Residential and commercial sectors 31

       2.3.2. Transport sector 43

       2.3.3. Industrial sector 50

       2.3.4. Efficient use of materials 59

       2.3.5. End-use and primary energy: the potential of efficient conversion 62

       2.3.6. Findings and international comparisons 68

    2.4 Regional and global implications 75

       2.4.1. IIASA "low" scenario 82

       2.4.2. The Columbo/Bernadine scenario 85

       2.4.3. An efficiency scenario 87

       2.4.4. Implications for the European Economic Community 94

3 WHAT KINDS OF ENERGY WILL WE NEED? 98
3.1. Matching heterogeneous end-use needs 98
3.2. Evolving thermodynamic structure of end-use needs 102

4 WHAT ENERGY SOURCES CAN WE SUBSTAINABLY USE? 104
4.1. Monolithic versus diverse/competitive strategies 104
4.2. Renewable sources with inherently low climatic risk 107
    4.2.1. Technical and commercial status 108
    4.2.2. Economic status at the margin 117
    4.2.3. National, regional, and global adequacy 119

5 BY WHAT POLICY INSTRUMENTS? 128
5.1. Strategies for implementation: conflicting philosophies 128
5.2. Price signals 129
5.3. Market imperfections 131
5.4. Accelerating retrofit or turnover of capital stocks 132

6 HOW QUICKLY CAN THESE THINGS BE DONE? 136
6.1. Rate and magnitude problems 136
6.2. Rate constraints and comparisons 140
6.3 Comparative rates: theory and practice 142
6.4 Oil displacement and other short-term imperatives 145

7 IMPLICATIONS FOR CLIMATIC RISK 149
7.1. The cost of failure 149
7.2. Conclusions 152

REFERENCES 165

About the authors 183

FIGURES

1.1 Approximate projections of atmospheric carbon dioxide levels as a function of the pattern of increase or decrease in the rate of burning fossil fuels, 1975–2025   10

2.1 Patterns of energy use in the Federal Republic of Germany, 1973   29

2.2 Cost-effective potential efficiency improvements in the FRG residential and commerical sectors, at constant 1973 activity levels   44

2.3	Cost-effective potential efficiency improvements in the FRG transport sector, at constant 1973 activity levels	51
2.4	Cost-effective potential efficiency improvements in the FRG industry, at constant 1973 activity levels, including effects of materials policy	58
2.5	Technical potential of cost-effective improvements in FRG end-use and primary demands at constant 1973 activity levels assuming all future end-use energy is converted from fossil fuels	63
2.6	A global energy-efficient scenario (by regions) with strong economic growth and constant urban friction	93
3.1	End-use structure of selected industrial countries, ca. 1975	101
7.1	Rate of CO_2 emission, CO_2 concentration, and global average temperature change (C°) for a variety of energy scenarios	160

TABLES

1.1	Carbon dioxide levels as a function of fossil-fuel burn trajectory	9
1.2	Possible future shifts in fossil-fuel mix and CO_2 coefficient	9
2.1	Energy use in the FRG, 1973	26
2.2	Structure of energy end-use in the FRG, 1973	26
2.3	Cost-effective electricity savings in household appliances	40
2.4	Cost-effective potential efficiency improvements in the FRG residential and commercial sectors	42
2.5	Cost-effective potential efficiency improvements in the FRG transport sector	50
2.6	Cost-effective potential efficiency improvements in FRG industry	57
2.7	Potential energy efficiency improvements from reducing excess and lost materials and from increasing recycling of materials	61
2.8	Technical potential of combined efficiency improvements in direct and indirect FRG industrial energy use	62
2.9	Cost-effective potential efficiency improvements in energy end-use in the FRG	62
2.10	Technical potential of cost-effective efficiency improvements in FRG primary energy use	67

v

TABLES (continued)

2.11	Evolution of approximate estimates of US primary energy demand in the year 2000	73
2.12	Two ways to make 200,000 metric tons of fixed nitrogen per year	77
2.13	IIASA population assumptions by region	83
2.14	Assumed growth of Gross Domestic Product in the IIASA [1978] "low" scenario	83
2.15	Illustrative impact of aggregated efficiency improvements on IIASA (1978) "low"-scenario projections of primary energy demand	84
2.16	Possible urban populations in 2030	88
2.17	Possible world primary energy demand (excluding feedstocks) with efficient use, strong economic growth, and constant urban fraction	92
2.18	Population assumptions for Western Europe within Region III	94
2.19	Possible future primary energy demand in the EEC assuming rapid economic growth and economically efficient energy use	96
3.1	Percentage of total delivered energy needed in various forms in selected industrial countries around 1975	100
3.2	Possible evolution of FRG end-use structure	103
4.1	Characteristics of the Saskatchewan Conservation House	115
4.2	Potential renewable supply calculated in selected national studies	120
4.3	Primary demand estimates from selected national studies, compared with calculations based on the FRG/Colombo & Bernadini results	121
4.4	Potential renewable supply fractions for selected countries	122
4.5	Preliminary, approximate Nordic renewable supply/demand balances	124
6.1	Some illustrative rate-and-magnitude calculations for displacing fossil fuels by central electrification	138
7.1	Global rate of primary energy consumption (TW) and supply sources for a variety of energy scenarios, 1975–2030	159

Foreword

STEPHEN H. SCHNEIDER
WALTER ORR ROBERTS

The potential seriousness of carbon dioxide-induced climatic changes has been at the forefront of our thinking, writing and speaking for many years.* Indeed, despite a number of large remaining uncertainties, the importance of the CO_2 problem has been underscored by a growing consensus of climate scientists. This is evidenced by a number of recent assessments of the CO_2 issue from such conservative, deliberative bodies as the World Meteorological Organization, the Climate Board of the U.S. National Academy of Sciences, and even the U.S. Department of Energy.** Thus, it was with a sense of great anticipation that we received news that *Least-Cost Energy: Solving the CO_2 Problem* had been written.

* For examples see Roberts, W. O., "The Costs of Climatic Impacts" in *Interactions of Energy and Climate,* W. Bach, J. Pankrath, & J. Williams (eds.), D. Reidel Publishing Company, Dordrecht, Holland, 1980, pp. 461–468; Schneider, S. H., with Mesirow, L. E., *The Genesis Strategy: Climate and Global Survival,* Plenum Publishing Corp., New York, 1976, 419 pp.; Schneider, S. H., "Comparative Risk Assessment of Energy Systems, *Energy,* 1980, *4,* 919–931; Schneider, S. H., "The CO_2 Problem: Are There Policy Implications—Yet?", *Climatic Change,* 1980, *2,* 203–205.

** *An Assessment of the Role of CO_2 on Climate Variations and Their Impact,"* Joint WMO-ICSU-UNEP Meeting of Experts, Villach, Austria, November 1980, World Meteorological Organization, Geneva, January 1981, 29 pp.; Ad Hoc Study Group on Carbon Dioxide and Climate, *Carbon Dioxide and Climate: A Scientific Assessment,* Climate Board, U.S. National Academy of Sciences, Washington DC, 1979, 25 pp.; Report of the CO_2/Climate Review Panel of the Joint Climate Board/Committee on Atmospheric Sciences, Climate Research Committee, National Academy of Sciences, Wash-

Somewhat surprisingly, though, before launching into an enthusiastic and detailed account of this new work, Amory Lovins and colleagues started to tell us about the book by first apologizing, saying, in effect: I hope you climatologists who spend so much of your time on CO_2-related research aren't too upset, since this book shows that the CO_2 problem need never occur! We later understood why, as right on the first page of the Executive Summary the authors state:

> Although the climatological community realizes that prevention of the CO_2 problem is easier than cure, that prevention involves the discipline of energy policy, not of climatology. Most CO_2 studies have simply *assumed* that the rate of burning fossil fuel must and will increase rapidly, continuously, and indefinitely. But recent discoveries in another discipline—the engineering/economic analysis of efficient energy use—cast grave doubt on that assumption, for if energy is used in a way that saves money, the rate of burning fossil fuel will not increase but decrease. This is because identical, and even greatly increased, energy services can still be provided using less total energy—by using it more productively—and also by replacing most or all of the remaining fossil-fuel supplies with renewable sources. These twin measures would use only technologies that are presently available, highly cost-effective, and indistinguishable in reliability from present energy systems.

Thus, the authors argue that far from being economically infeasible, their strategy for avoiding the CO_2 problem (and others such as acid rain) is the only thing the world can afford; it is the conventionally projected *high*-energy futures, they contend, which are not economically feasible! It may seem then that we climatologists needn't bother to build more elaborate computer models of air/sea/ice interactions in order to improve estimates of the global temperature or regional precipitation changes from a CO_2 doubling if no such doubling will occur. Moreover, why should we worry about shifting grainbelts or potential agricultural changes or other human impacts? Neither

---

ington, DC, 1981 (in preparation); *Environmental and Societal Consequences of a Possible CO_2-Induced Climate Change: Our Research Agenda, Vol. 1,* No. 013 of Carbon Dioxide Effects Research and Assessment Program, DOE/EV/10019-10, U.S. Department of Energy, Washington DC, December 1980.

should we, it might seem, worry too much more about how many people could be displaced or how many trillions of dollars of property lost by a CO_2-induced sea level rise from a deglaciation event, if no significant CO_2 increase from fossil fuel consumption occurs.

Despite the authors' best efforts, we don't fear for our jobs; nor do we resent their conclusions. Quite the contrary! We sincerely hope that the authors are accurate and complete in their analyses and conclusions as well as articulate and effective in spreading their views to the world—something they have an impressive record for doing in the past.

Our reasons for this attitude are several. The authors make no claim that CO_2 research should be abandoned; indeed it was such research which impelled them to re-analyze whether the energy policy which would commit us to a doubling of CO_2 is even the best way to meet our energy needs. From a climatologist's point of view, it would be the best of both worlds if the authors' least-cost energy strategy would also minimize a potentially serious environmental problem. But we don't believe research on CO_2 effects and impacts will—or even should—be slowed because of *Least-Cost Energy: Solving the CO_2 Problem*. That is because CO_2 is such a basic part of climatic theory that much more work on it is needed to help explain past climates like the ice ages, or present climates on other planets like Mars and Venus. Moreover, the potential societal impacts and responses to the advent or prospect of increasing CO_2 are closely related to other basic environment/society problems like drought-induced stress, acid rain, and long-term soil erosion. Such "tie-ins" assure that CO_2-related research will have spinoffs to other problems, regardless of the ultimate seriousness of CO_2 *per se*.

Furthermore, other trace gases like chlorofluorocarbons and nitrogen oxides are building up in the atmosphere, and can cause an enhanced "greenhouse effect" like that for CO_2. In this connection, the authors examine only fossil-fuel-based causes of CO_2 increase. Deforestation and other land uses could also cause a significant—but ultimately much smaller—CO_2 buildup, even if fossil-fuel use is largely curtailed. And finally, there always is the possibility (remote, we hope!) that these analyses could be flawed or (much more likely) that people will be characteristically too slow to change their old energy habits, despite the strong challenge and logic in *Least-Cost Energy: Solving the CO_2 Problem*. For all these reasons, and because we sincerely hope that this work stimulates a far-reaching debate on energy-related planning, we are delighted both to write this Foreword and to commend the book to all thoughtful readers concerned with the world's future. Their actions—your actions—can help make a number of the possibilities outlined here a global reality.

One of the features we appreciate most in this work is the high degree of quantitative argument used to support its analyses. Moreover, by stressing the "least-cost" approach based on current technology, the book should appeal to "bottom line"-oriented managers and decision makers. The authors stress the "bottom-up" approach in their analyses of the potential economic power of billions of individual and community decision makers responding to market signals, rather than the more conventional "top-down" approach so often used to make projections of future energy demand—the approach that has produced the high-CO_2-risk scenarios so prevalent today. Furthermore, the often-held pessimistic view that nothing can be done to prevent the CO_2 buildup, so we'll merely have to adapt to it as it unfolds*, is in large part a consequence of such "top-down" extrapolations of energy use. Thus, we welcome the fresh approach of Lovins, Lovins, Krause and Bach to energy demand scenario-building in general, and to CO_2-minimizing aspects in particular.

We may personally differ—both with each other and with the authors—over some issues, particularly the relative desirability of various alternative energy supply options. Nonetheless, we strongly endorse their thoughtful attempt to put the foundation of the CO_2 problem—energy use projections—back into the forefront by making *explicit* the various behavioral assumptions and value judgments which underlie projections of future energy supply options and end uses. As climatologists we know that our

---

* Meyer-Abich, K. M., "Socioeconomic Impacts of CO_2-induced Climatic Changes and the Comparative Chances of Alternative Political Responses—Prevention, Compensation, and Adaptation," *Climatic Change, 2,* 373-385.

leading colleagues will read and be influenced by *Least-Cost Energy: Solving the CO$_2$ Problem*. We can only hope that the book attracts a far broader audience and sparks the kind of fundamental debate and rethinking so desperately needed if energy/environment problems are to be solved.

STEPHEN H. SCHNEIDER*
National Center for Atmospheric Research**
P.O. Box 3000
Boulder, Colorado 80307

WALTER ORR ROBERTS*
Aspen Institute for Humanistic Studies
  and
University Corporation for Atmospheric Research
Boulder, Colorado 80307

* The National Center for Atmospheric Research is sponsored by the National Science Foundation.

** Any opinions, findings and conclusions or recommendations expressed herein are those of the authors and do not necessarily reflect the views of the National Science Foundation.

Preface

This book grew out of the concepts, data, and citations assembled in a short paper (Lovins 1980) rethinking the effect of world energy needs on global climate. Commissioned by Professor Wilfrid Bach and presented at a March 1980 international conference which he chaired, the paper suggested that burning fossil fuels faster and faster was far from inevitable or inexorable: indeed, it was not even an economically attractive route to prosperity. If this was true, then a good deal of climatological research—as well as the major thrust of world energy policy—had been based on mistaken premises about the economics of using energy. This preliminary finding clearly called for a more detailed analysis to test, quantify, and extend the results on a national, regional, and global scale.

Professor Bach undertook to seek official support for such a joint research project. His success enabled us to explore in depth the relationship between energy, economics, and climate. The results far surpassed our expectations, largely because of our good fortune in our collaborators. Although we carried most of the burden of formulation and writing, we also secured the cooperation of our friend and colleague, Dr. Florentin Krause. Earlier the lead analyst and author of the first detailed study of an economically efficient energy future for the Federal Republic of Germany, Dr. Krause was largely responsible for the centerpiece of this joint analysis: a case-study of energy efficiency in the

FRG (Section 2.3) and its extension to Europe and the world (Section 2.4). Professor Bach contributed a vivid computer analysis (Figure 7.1) of the climatic implications of our global "efficiency scenario," as well as helpful guidance and criticisms throughout.

This international collaboration produced a variety of products:

- a detailed, up-to-date case-study of how much energy efficiency is economically worth buying in the FRG—already among the world's most energy-efficient countries. We showed that cost-effective technologies now available could reduce by at least 82% the energy needed to produce a unit of German GNP (compared to the base year 1973).

- an extension of this "efficiency scenario" to a regional (European) and global scale. These extrapolations were cross-checked for consistency with previous national studies and then integrated with analyses of the cost-effective role of presently available renewable energy sources.

- a critical comparison of these results with those of a widely publicized study by the International Institute for Applied Systems Analysis (IIASA). That study purported to show that vast increases in energy use, and particularly in the use of coal and nuclear power, are essential for global development. We found that the IIASA results are an artifact of failing to consider carefully either the state of the art of efficiency-improving and renewable energy technologies or their major cost advantages.

- a wide-ranging discussion of recent developments in energy policy, economics, politics, analytic methods, and technologies, with special emphasis on practical ways to raise energy efficiency and harness renewable sources.

- an analysis of the principles and the quantitative pathways by which an energy strategy based solely on minimizing direct economic costs to consumers can at the same time avert global climatic changes (notably the "greenhouse effect") arising from the burning of fossil fuel.

This book, then, should be useful not only to climatologists but also to economists and to all those concerned, professionally or as informed citizens, with energy policy. Furthermore, since the same measures we describe for minimizing the release of carbon dioxide into the air would also minimize other impacts of using fossil fuels (such as acid rain), we hope our results will help readers with a wide range of energy-related concerns.

Our use of a German case-study to support our wider analysis is something of an historical accident, but we hope it offers several unexpected side-benefits. First, it can help to introduce non-Europeans—and perhaps some Europeans too—to a valuable body of national energy literature that is too seldom exported. Although the FRG is not nearly as energy-efficient as would be economically worthwhile, it is still far more energy-efficient than, say, the United States, and knows a lot about how to get there. Second, while we focus on a heavily industrialized, economically successful country, we consider also the need for long-term, sustainable development in the poorer countries within a quite different cultural setting. The German experience has a good deal to contribute, we feel, to this international and global perspective. Third, by synthesizing the best German literature with what we believe are the best up-to-date data on efficiency and renewables from many other countries, we have sought to portray the rapidly evolving state of the art in the most technically advanced nations more fully than we could by relying on data from any one nation. For reasons we discuss at length, the FRG is arguably the least favorable case we could have picked for our efficiency-and-renewables thesis: virtually any other country should be able to do the same things more easily.

For their enterprise, professionalism, and hard work in helping us to bring this good news into print in a few short months, we are much indebted to our publisher, Jack Howell, our production manager, Mike Fender, and their colleagues at Brick House and its contractors. Our valued colleague Roger Sant, to whose work this book owes much inspiration, graciously consented to our use of "least-cost" and "least-cost energy strategy," both of which are registered trademarks of his new enterprise, Applied Energy Services, Inc. Finally, for her patient and generous hospitality during our harried sessions with word processor, typewriter, calculator, telephone, and other infernal machines, we are all deeply grateful to Farley Hunter Sheldon.

Woody Creek, Colorado

AMORY B. LOVINS
L. HUNTER LOVINS

Executive Summary

CHAPTER 1. INTRODUCTION
1.1. Energy/climate risks: a future problem whose time has passed?

The present rate at which people use nonsolar energy—about 9 TW (9×10^{12} watts)—causes minor, localized disturbances to the earth's climate. But about 8 of the 9 TW is derived from burning fossil fuels. This releases various pollutants, notably carbon dioxide gas, about half of which accumulates in the atmosphere. Industrial activity has already increased the CO_2 level by about a fifth, and is commonly projected to cause a doubling by about the middle of the next century. Although there are still many uncertainties, most climatologists believe that a doubling of the CO_2 level would warm the earth's surface by an average of several Celsius degrees—several times more at the Poles—and that this warming would cause very disruptive changes in weather patterns, including distribution of rainfall, and perhaps in sea level. These changes would probably be irreversible. No practical way is known to prevent them—if the CO_2 level does rise to such levels as to cause them.

Although the climatological community realizes that prevention of the CO_2 problem is easier than cure, that prevention involves the discipline of energy policy, not of climatology. Most CO_2 studies have simply *assumed* that the rate of burning fossil fuel must and will increase rapidly, continuously, and indefinitely. But recent discoveries in another discipline—the engineering/economic analysis of efficient energy use—cast grave doubt on that assumption, for if energy is used in a way that saves money, the rate of burning fossil fuel will not increase but decrease. This is because identical, and even greatly increased, energy services can still be provided using less total energy—by using it more productively—and also by replacing most or all of the remaining fossil-fuel supplies with renewable sources. These twin measures would use only technologies that are presently available, highly cost-effective, and indistinguishable in reliability and convenience from present energy systems.

These findings, expressed in careful, detailed, and highly sophisticated analyses from more than a dozen countries, are just now becoming widely known in the energy policy community, but are not yet known to most climatologists. This study surveys the extent, cost, and climatic implications of such major fossil-fuel savings. Those savings are so large that energy/climate problems may well be completely preventable, and at worst could be postponed for additional decades or centuries, providing ample time to achieve fuller use of energy alternatives. Best of all, the energy policy measures which can achieve this low climatic risk also make sense on other grounds (especially economics) and should therefore be done anyhow.

1.2. How much time can we buy?

Quantitative estimates of when various CO_2 levels would be reached, as a function of the future pattern of burning fossil fuel, show that the CO_2 problem is a matter not of fate but of choice. Energy policy can determine with enormous flexibility when troublesome levels will arise. Given the uncertainty in what levels might be dangerous, there is a strong argument that the global rate of burning fossil fuel should at least not rise any further. The usually forecast warming would probably then be only half as big and occur a half-century later. But if instead the rate of burning fossil fuel is actually reduced, then significant climatic changes will be greatly delayed or even prevented altogether. The amount of delay will be disproportionately larger than the reduction: it will be the product of the size of the reduction times how soon it occurs. How great a reduction, then, can be achieved how soon? This book seeks to answer that question.

CHAPTER 2. HOW MUCH ENERGY WILL WE NEED?

Neither economic nor engineering analysis by itself can reveal practical energy futures. But a careful combination of both techniques, with due attention to social and psychological conditions, can clearly show what is both possible and worth doing. This book uses engineering economics to analyze the results of following a "least-cost energy strategy"—one that seeks to provide needed energy services at the lowest possible direct economic cost, as would occur in a competitive free market. To allow for uncertainties, the analysis uses many "conservatisms"—assumptions deliberately biased in the direction least favorable to the conclusions. The calculations rest on documented empirical cost and performance data, and are designed to be easily understood, reproduced, or modified. The scope of the analysis is national (including the Federal Republic of Germany, or FRG), regional (the European Economic Community, or EEC), and global. Its time-scale is 50 years—about as long as major energy facilities ordered today are expected to work—and several decades beyond. The scenarios presented are not forecasts of what *will* happen, but rather illustrations of what *could* happen under an economically efficient energy policy.

2.1. Defects of conventional forecasts

Most climatological analyses assume that by 2030, people will use energy at a rate not of about 9 TW as now but of about 30–80 TW. The International Institute for Applied Systems Analysis (IIASA) has recently published "low" and "high" scenarios reaching 22 and 36 TW respectively in 2030, and the lowest value given by a major study is 16 TW, by Colombo and Bernadini (whose methodology we shall use later). Our analysis, by using the best available evidence about cost-effective technical improvements in the efficiency of using energy, derives world energy needs of only 5–8 TW, and perhaps less than 4 TW, for an equally populous and prosperous world.

Previous analyses envisaged long-term world energy needs two to ten times higher than we do because they did not properly consider most or all of twelve effects: the tendency of higher price to elicit greater efficiency; the dependence of historically high energy growth rates on historically declining real energy prices; subsidies which have reduced apparent energy prices; saturations in energy service needs; past promotional policies; distortions in the historic data used as a basis for forecasts; structural changes in the national and world economy; improper allowance for energy conversion and distribution losses, which have often been treated as a part of demand rather than derived from it; the difficulty of adding up many small terms of energy saving if the study is too aggregated to detect most of them; rapid technological progress in energy productivity and alternative sources; the need to apply the same tests to all technologies considered; and the need to compare with each other all opportunities for providing additional energy services. These deficiencies, which are prominent in the IIASA and other scenarios showing high energy demand, are systematically identified and avoided in our analysis.

2.2. Need for improved forecasts

Both sound national energy policy and a correct assessment of climatic risk require that energy needs be analyzed in great detail, taking account of engineering constraints and rigorously applying tests of cost-effectiveness. In the past few years, such studies have been done, mainly in the private sector, in and for a wide range of industrial countries. They show how to use straightforward, economically attractive, presently available "technical fixes" to decrease by a factor of three to six or more the amount of primary energy needed to produce a unit of economic activity. Some of these several dozen "efficiency scenarios" are in fact the most detailed studies of energy demand so far done anywhere—especially those for the Federal Republic of Germany, United Kingdom, United States, and Denmark.

2.3. Case study for the Federal Republic of Germany

The Federal Republic has a diverse economy dominated by heavy industry; a rigorous climate; a relatively high present level of energy efficiency; a slight net export of "indirect" energy embodied in goods and services; and a relatively slow rate of diluting or replacing the present stocks of buildings, vehicles, factories, and equipment with more efficient versions. These characteristics make the FRG an ideal subject for a case study to be used as an existence proof or *a fortiori* argument for other countries, because such an extrapolation will virtually always exaggerate the amount of energy which those other countries would need.

Our analysis considers 15 sectors of German energy use in 1973—a convenient base year because it has good statistics and predates recent efficiency improvements, thus avoiding double-counting. By citing the extensive German and international literature on technologies which are currently in or entering commercial service, and which are cost-effective against present or marginal energy prices, we show for each of these sectors how much its specific energy intensity—the energy needed to produce a unit of good or service—can practically be reduced. Where necessary, we distinguish the extent to which these measures can be readily deployed over the next 50 years or so (which we call the "2030" case) from the additional savings that can be achieved, especially in buildings, by waiting a little longer so that the existing stock can be completely replaced (the "present technical limit" case). In each instance we understate the amount and speed of the achievable energy saving, and we omit any potential saving resulting from changes in values or lifestyles. It is also unlikely that any of our proposed investments in energy efficiency goes as far as would be economically worthwhile compared to increasing the supply of energy.

2.3.1. RESIDENTIAL AND COMMERCIAL SECTORS

Proven methods of building construction and alteration now in routine use by contractors in several countries, with typical payback times of a few years at present energy prices, can reduce the net energy required for comfortable heating of the entire interior of new German houses by at least 95%, and of old houses by at least 80%, below the energy required for space heating in 1973 (when most houses were only partially heated). The space heat saving in commercial buildings is probably as large, but since the data are less complete we assume a smaller saving. At least half of the heat thrown away in used hot water can be straightforwardly recovered to preheat fresh hot water. At least 70% of the heat used for other processes in buildings—mainly greenhouses, some cooking—can also be saved by proper engineering design or modification. At least two-thirds of the electricity used in household appliances (assuming that each household has every appliance) can be saved by good design, with a present FRG payback time under five years. Recent Canadian and Swedish office buildings show a similar fraction saved and a similar payback time for lighting and other electrical functions in commercial buildings. Collectively, the specific measures we document would reduce the total energy use of FRG office buildings (starting at 1973 levels of efficiency) by an average of 82% over the next fifty years or so, and by a total of 90% once the old buildings (except historic ones) have all been replaced. Buildings accounted for 43% of 1973 FRG final energy use.

2.3.2. Transport sector

Another 19% of that energy use was for transport—over half of that for cars, which in 1973 had an average efficiency of 10.6 ℓ/100 km (22.5 mi/US gal). Many feasible combinations of measures to optimize power trains, reduce weight and friction, and streamline cars can improve that efficiency by at least 77%, to 2.4 ℓ/100 km (100 mi/US gal) for four-passenger cars: in fact, an advanced Golf (Rabbit) prototype from VW was recently tested in the US at 3.0–2.4 ℓ/100 km (80–100 mi/US gal) on the EPA city/highway circuit. (Increasing the fleet fraction of two-passenger cars—which could equally well make most FRG trips—would increase the fleet-average efficiency to 1.9 ℓ/100 km or 125 mi/US gal.) The economics of such improvements are excellent: an improvement even to 4 ℓ/100 km (60 mi/US gal) would save the average German driver, at present gasoline prices, about $545 per year, repaying the extra capital cost of the car in two or three years.

Broadly similar opportunities apply to trucks and buses: the most cost-effective measures for improving the vehicles and their load management would reduce their energy use per ton-km by

60%. At least 25% can be saved by better design of railways and ships (a Japanese ship has cost-effectively saved over 50%). New aircraft already being introduced into the fleet will improve its average efficiency by half, with even further gains in prospect. Combined with the savings in cars, these cost-effective technical measures would reduce German energy requirements per unit of transport by 64% in the "2030" case and ultimately by a total of 68% (conservatively assuming no additional long-term improvement except in cars and aircraft).

2.3.3. INDUSTRIAL SECTOR

Of the 38% of German final energy used by industry in 1973, over half was used to make steel, chemicals, and cement. Standard technologies which are in any case necessary to keep the German steel industry competitive reduce energy requirements per ton by 23% for fuel and 35% for electricity, or a total of 40% in primary energy after waste heat and by-product fuel gases are harnessed. Similarly straightforward innovations already being rapidly introduced in the chemical industry will save about half its total energy, and likewise at least a third in the cement industry. In industries outside these "big three," proper sizing, coupling, and controls on electric motors cost-effectively save at least a third of their electrical needs, as do available improvements in aluminum smelters. Insulation, heat recovery, cogeneration, heat pumps, electronic process controls, and other well-known technical measures can cost-effectively save half the process heat. The saving in space heat is about the same in factories as in office buildings. Together, these measures save in the "2030" case 47% of the total direct fuel and 36% of the electricity that German industry used in 1973.

2.3.4. EFFICIENT USE OF MATERIALS

Industry uses energy to produce and fabricate materials. Much of this energy (plus imported raw materials) can be conserved or recovered by cost-effective measures for recycling materials; recycling products by remanufacture, reworking, or reuse; reducing dissipative uses; reducing the material removed in milling and manufacturing; reducing post-consumer waste; eliminating unnecessary materials in products; and increasing product lifetimes. The technical potential for such improvements has been carefully surveyed in the FRG. Presently commercial technologies to recycle materials and to reduce excess materials in manufacturing can save respectively 7% and 16% of 1973 German industrial energy needs, or in combination, 22% (not counting the saving in petrochemical feedstocks, which we account for separately later). More systematic and sophisticated measures would raise this to at least 50%. This indirect energy saving combines with the direct ones described above for industry to yield a 57% total saving of industrial energy in the "2030" case and 72% ultimately, even assuming no additional long-term gains in direct energy efficiency in industry.

2.3.5. END-USE AND PRIMARY ENERGY: THE POTENTIAL OF EFFICIENT CONVERSION

The measures so far described would reduce the FRG's total delivered energy needs, compared with 1973 levels of technical efficiency, by 69% in the "2030" case and 79% in the "present technical limit" case. How much primary energy would be needed to deliver that much final energy? The maximum value would result if all energy came from fossil fuels: this ignores even the present German hydroelectric capacity, which would meet 17% of the "2030" total electrical demand. Analysis of the cost-effective potential for combining the production of heat and electricity in the FRG, using presently commercial equipment, shows that the total conversion efficiency of primary to end-use energy can be improved from 0.73 in 1973 to 0.87 in the "2030" case. Thus, combining the efficiencies documented for each end-use sector and for the energy supply industries themselves, the cost-effective use of best available technologies would reduce total primary FRG energy demand (including feedstocks) by 70% in the "2030" case and by 80% in the "present technical limit" case. These two cases thus entail respective end-use energy demands of 1.18 and 0.80 kilowatts per person (kW/cap); total primary fuel demand (excluding feedstocks) of 1.36 and 0.92 kW/cap; and total primary energy demand (including feedstocks) of 1.72 and 1.15 kW/cap, compared with the actual 1973 value of 5.67.

2.3.6. Findings and international comparisons

This finding that the total primary energy productivity of the 1973 FRG economy can be cost-effectively increased by 3.3- to 5.0-fold with no new technologies is consistent with similar analyses in other industrial countries. Extremely detailed European and North American studies, both official and private, have shown comparable or larger savings, after due allowance for differences in initial efficiency. Two authoritative US studies have recently confirmed, as earlier European analyses implied, that the efficiency gains proposed are so economically advantageous that they would *reduce* the fraction of future GNP used to buy energy services—so that far from driving inflation, the energy sector would become a net exporter of capital to the rest of the economy! As efficiency scenarios become more disaggregated, localized, and alert to new technical advances, they are showing ever larger opportunities for saving energy.

So far we have considered only how much energy would have been needed to run the 1973 FRG economy at cost-effective levels of technical efficiency. Other analyses which we cite have applied similar savings to a future pattern of continued industrial and economic growth, and thus calculated possible future patterns of energy demand for the FRG and other major industrial nations. Even the most generous assumptions about economic growth cannot overcome the severalfold efficiency improvement that results simply from investing to provide energy services at least cost. Accordingly, if the industrialized nations are economically efficient, their energy needs can be expected to go down, not up.

2.4. Regional and global implications

These conclusions can be extended to a regional or global scale by analogy, by scoping calculation, or by explicit aggregation of the main terms. We next apply a combination of these methods.

Reasonable demographic and economic assumptions for all Western and East-bloc industrial nations, combined with efficiency improvements smaller than those just described for the FRG, would decrease those countries' total energy needs by about 40% over the next 50 years. To determine whether this conclusion would be reversed on a global scale if the aspirations of the developing countries were also satisfied, we consider two approaches. First, for the sake of argument we assume that all developing countries will follow precisely the same development path that the OECD countries followed under historically unique conditions—so far that the whole world will look like the FRG looked in 1973. Assuming this complete heavy industrialization to be possible and desirable, and assuming in 2030 a population of 8 billion, this scenario would imply total world primary energy use somewhat under 8 TW—that is, *less than the present level*. This is because the same levels of technical efficiency that the FRG can attain can also be attained elsewhere, but faster and cheaper, because the developing nations can build their energy-using infrastructure efficiently the first time rather than needing slow and costly "retrofits" later. This 8-TW *Gedankenexperiment*[1] is conservative in many respects—not only economically and technically, but because it caricatures a concept of development which has not proven successful and has been largely abandoned by the responsible international agencies. Yet it suggests that humankind could, in principle, avoid the potential climatic impact of prolonged and increasing fossil-fuel use without limiting worldwide material standards to less than the current level of highly industrialized European societies.

As a more practical alternative to this hypothetical world of "two billion Chinese driving Buicks," we consider also a future course of development nearer the ideal expressed by such bodies as the Brandt Commission. To represent diversity better, we also disaggregate the world into the seven regions used in the IIASA study. And to capture the important difference between the changes in the energy intensity of providing energy services (the technical efficiencies described earlier for the FRG) and the changes in the amount of energy services composing a unit of GDP (a function of the stage and pattern of development), we also distinguish these two factors and subject them to separate calculations.

---

1. That is, an experiment carried out in theory rather than in practice; a thought-experiment.

We first apply an efficiency correction to the "low" scenario[2] of the IIASA study by assuming its regional economic growth factors (which increase the Gross World Product by a factor of 3.69 from 1975 to 2030) but our model FRG efficiency improvements. The resulting world primary energy demand in 2030 is 32% below its 1975 level, rather than 235% above it as the IIASA group calculated. But this does not yet account for the application of modern development principles. To do this we take advantage of a methodology developed by Umberto Colombo, the President of the Italian Atomic Energy Commission, and his colleague Bernadini, in their research for the EEC. They have analyzed in detail the energy effects of a less centralized development pattern emphasizing the provision of jobs, food, and other necessities where people already are, rather than only in vast urban areas. This approach seems more consistent with social and biological constraints, with modern emphasis on more equitable land tenure, with an appreciation of the unique historic conditions that led to the urbanization of the European peasant, and with a pragmatic assessment of how best to use limited development resources. Its shorter travel and haulage distances, materials savings, greater local self-reliance, and more biologically sensitive farming and forestry methods would together, according to the Colombo & Bernadini case-study, reduce total energy requirements 2.8-fold compared with a highly urbanized settlement pattern. Their scenario for 2030 assumes that the fraction of world population living in cities will then be the same as it is today (ca. 30%). Because this fraction does not greatly increase, as most analyses assume, the average energy/GDP ratio decreases by 30%, so world primary energy demand in 2030 is only 16 TW. But this result does not take careful account of cost-effective improvements in the technical efficiency of using energy.

We therefore combine these approaches in an assumed world that stabilizes its population at 8 billion in 2030. We assume the IIASA "low" scenario economic growth rates to 2030 (by which time the poor countries are 10 times as rich as they are today, but

2. Our analysis uses mainly 1978–79 IIASA demands 15% above the final (1981) data.

still ten times poorer than the rich countries per capita). For the additional period 2030–2080 we assume 1.1%/y GDP growth in the developing countries and none in the presently industrialized countries, which by then have average per capita GDP of over $10,900 (in 1975 dollars). Applying to such a world our "2030" efficiency coefficients for 2030, our "present technical limit" efficiencies for 2080, and the Colombo & Bernadini pattern of global development, we then calculate—with implicit conservatisms—a total world primary energy demand of about 7.1 TW in 2000, 5.2 TW in 2030, and 3.6 TW in 2080. This last figure is less than half the present level despite the assumed 4.6-fold increase in Gross World Product. The combined effect of assuming a more appropriate development path and an economically efficient level of energy productivity is to make our world energy demand in 2030 five times lower than IIASA's so-called "low" scenario. Thus analysis more refined and disaggregated than our earlier 8-TW *Gedankenexperiment* re-emphasizes that most published estimates of long-term world energy needs have barely begun to take advantage of financially and socially attractive opportunities for meeting human needs with an elegant economy of means.

For further specificity, we apply these methods to the European Economic Community. Using IIASA's population projections, GDP/cap values from Colombo & Bernadini (but in close agreement with those of the IIASA "low" scenario and of national studies), and the energy intensity coefficients of our FRG model (already among the most energy-efficient countries in Europe), we obtain a 2.4-fold increase in economic activity by 2030 as compared with 1975, but simultaneously a 57% reduction in EEC primary energy needs—coincidentally the same size as the EEC's current 55% dependence on imported energy.

CHAPTER 3. WHAT KINDS OF ENERGY WILL WE NEED?

3.1. Matching heterogeneous end-use needs

To understand the climatic relevance of future energy needs, we must consider not only quantity but also quality. The energy problem is not simply where to get more energy, of any kind, from any source, at any price. There are different forms of energy whose different prices and qualities suit them to different uses. Further, it is not energy *per se* that we require, but rather the services that energy gives us, such as comfort, light, mobility, and smelting. It is therefore more sensible to view the problem as how to provide the *amount, type, and source of energy that will do each desired task in the cheapest way.*

Viewed in this light, the tasks for which energy is currently required at the point of final use in the FRG (which is fairly typical of industrial countries in this respect) are 75% for heat (two-thirds of which is needed at temperatures below 100°C), 18% for portable liquid fuels to run vehicles, and only 7% for the premium applications which need electricity and which can use this very costly form of energy to economic advantage. In no industrial country can additional electricity be used cost-effectively, because the "electricity-specific" needs are already met by present capacity with a good deal left over. In contrast, most energy forecasts *assume* additional rapid electrification which could only be used for heating (and ultimately perhaps for electric cars). That assumption cannot withstand economic scrutiny. Yet more than any other single factor, it is responsible for the very high forecasts of primary energy demand which have rightly caused so much climatic concern.

3.2. Evolving thermodynamic structure of end-use needs

Cost-effective improvements in FRG energy efficiency would apply evenly enough to the different final forms of energy requirements that their proportions would not change dramatically over the next century or so. The share required as heat would drop from 74% to about 55%, but half the heat would still be at low temperatures. The share of vehicular liquid fuels would rise from 19% to about 28%, and that of electricity from 7% to about 17%, but the absolute amount of both would decline markedly. These patterns, which generally correspond to the least-cost forms of energy supply, do not imply major shifts in the long-term shape of the energy system.

CHAPTER 4. WHAT ENERGY SOURCES CAN WE SUSTAINABLY USE?

4.1. Monolithic versus diverse/competitive strategies

Of all options for replacing fossil fuels, only renewable sources can be essentially free of significant impact on climate. Non-renewable, monolithic energy systems, using small numbers of similar large machines, are easy to model, and most models of the future energy system cannot cope with anything else. In contrast, renewable energy sources are immensely diverse, and would probably come in large numbers and (for the most part) relatively small sizes, because matching their scale to that of the mainly dispersed uses can reduce total costs. This diversity is an advantage in cost and convenience, but complicates proper analysis of the renewable potential.

The state of the art has nevertheless been carefully studied for renewable sources providing heat at all temperatures, cooling, liquid fuels, and electricity. It shows that enough renewable systems are now proven and available to meet virtually all long-term energy needs (at cost-effective efficiency levels) in every industrial country studied, including the US, Canada, FRG, UK, France, Sweden, Denmark, and Japan. This list, containing countries that are variously cold, cloudy, densely populated, and heavily industrialized, suggests similar opportunities for most other countries where conditions are more favorable. Renewable supply looks even more attractive at either a more regionalized or a more localized level.

4.2. Renewable sources with inherently low climatic risk

4.2.1. TECHNICAL AND COMMERCIAL STATUS *and*
4.2.2. ECONOMIC STATUS AT THE MARGIN

Our analysis of renewable sources includes passive and active solar heating; passive solar cooling; solar process heat; conversion of farm and forestry wastes (but not, in general, of special crops) to alcohols and pyrolysates (oils made by heating woody matter with little air); present and small-scale hydroelectric power; windpower; and in some instances solar cells (photovoltaics). We do not assume geothermal energy (which is not generally renewable), tidal or wavepower, solar-thermal-electric conversion ("power towers"), ocean-thermal-electric conversion, biomass plantations, or solar power satellites, as we have not found a case where any of these last four technologies is economic or necessary. We also confine ourselves to sources which are relatively understandable to the user—although they can be technically very sophisticated—and which supply energy of appropriate quality and scale to do their task at least cost.

Correctly analyzing the potential of these "soft energy technologies" requires care, because the technological evolution is extremely rapid, much of the information is outside official channels, the devices and processes are very diverse, and their cost and performance depend sensitively on their design simplicity, market structure, scale, and efficiency of end-use. Nonetheless, several very thorough international studies have shown that the best available sources, performing at levels reasonably attainable in normal commercial practice, and cost-effective at least against the prices of non-renewable replacements for oil, are adequate for virtually all tasks, essentially independent of climate. (We consider climates less favorable than that of the FRG.) Many of these technologies are already diffusing into the marketplace. They are also being rapidly improved in both price and performance, although we ignore any possible future developments.

The storage of direct solar energy is not a serious problem if energy use is first made efficient. (We illustrate this by describing a "superinsulated" Canadian house whose solar heating system is large enough to cover all space and water heating needs without backup, even though its collectors have less than 10% of the floor area and its storage tank less than 3% of the volume of the house—five or ten times less than most studies say is needed for partial coverage in an inefficient house.) Electrical storage would likewise be a far less serious problem than in a conventional energy system, because with total demand reduced by efficiency gains, the grid would be dominated by present and small-scale hydroelectricity which can store water behind a dam. Indeed, the tendency of hydroelectricity, windpower, and solar cells to work best at different times and in different weather patterns would actually increase the reliability of supply. Finally, with efficient use, a completely "soft" energy system would generally require little or no extra land, even in cities, and would certainly need less land than conventional supply systems.

If these factors are properly considered—that is, if a renewable energy system is intelligently designed and efficiently used according to normal criteria of economic rationality—then presently available soft technologies are not only adequate but economically preferable for energy supply. Although not cheap, they are cheaper than non-renewable alternatives to replace the dwindling oil and gas: cheaper even in capital cost, and several times cheaper in delivered price of the energy service. This conclusion, which appears consistently in careful assessments in a wide range of conditions, is becoming more widely accepted as critics verify the references. It is, moreover, conservative in many respects, including its omission of "external" benefits of soft technologies—for employment, fast oil savings, monetary stability, environmental protection, nonproliferation, political convenience, Third World development, public and occupational safety, protection against technical failure, greater resilience in the face of surprises, and of course lower climatic risks.

4.2.3. NATIONAL, REGIONAL, AND GLOBAL ADEQUACY

Although the technical and economic potential of indigenous renewable sources has been well studied in most major industrial countries, those studies seldom assume that energy efficiency has first been raised to economically worthwhile levels. If this is done, renewable supplies become more than adequate even in countries supposedly poor in energy, such as France and Japan: they are poor in fuels but singularly rich in renewable energy, and with a cost-effective renewable supply system, both could be substantial net exporters of energy, as could Europe as a whole.

With lower levels of efficiency than our "present technical limit" case, and assuming a 2.4-fold increase from the 1975 GDP, there seems at first to be a slight deficit in trying to cover *all* long-term energy needs with present renewable sources in the two most difficult countries—the FRG and the UK. This deficit is in high-temperature process heat, chiefly for the steel industry, and can be overcome, even for the very large steel industry assumed, by any combination of steel/fuel coproduction from charcoal (currently considered promising in Sweden), additional wind or photovoltaic capacity delivering electricity or hydrogen, high-concentration-ratio solar collectors (probably economic in central Europe at present oil prices), or modest international energy trade. Assessments of the potential renewable energy interchanges among the Nordic countries and within Europe show that this approach may prove highly attractive. In the long term, many European countries can be net exporters of renewable energy to neighbors who might find trade more convenient than strict energy self-sufficiency, even though both approaches are feasible.

The potential renewable energy flows between the seven IIASA regions of the world have also been examined by Bent Sørensen, who conservatively assumed a global energy demand of 40 TW (at least five to ten times the value we found was economically efficient), including large amounts of high-quality energy. Nonetheless, after considering physical, biological, and land-use constraints, he found that each region of the world could be self-sufficient in renewable energy in each category, except for a small deficit in Asia, readily filled by exports from surplus regions. With more efficient energy use, that deficit would not arise, and the whole renewable supply system would become considerably more attractive. Parallel recent studies at IIASA find a global soft-technology potential of 15–34 TW (plus a central-electric solar potential of an additional 57–244 TW)—both so large compared to our long-term global energy needs of <4–8 TW that it appears the problem with renewable energy supply may be not that it offers too few attractive options but that it offers too many. Finally, by analyzing extreme climates—Stockholm and the Kenyan Highlands—T.B. Taylor has shown that attractive solar performance can be expected in both areas, and therefore by inference in between (where over 99% of the world's people live), even in areas not yet specifically studied.

The foregoing arguments about whether *all* long-term needs can be supplied by cost-effective and presently available renewable sources go much further than is necessary for our argument about climatic risk, for from this point of view it does not matter whether fossil fuel is replaced by efficiency improvements or by renewables. Although either will reduce CO_2 releases just as well, however, the former are generally cheaper, faster, and less controversial, and are the mainstay of our thesis that an economically efficient energy policy, nationally and globally, can largely or wholly eliminate the burning of fossil fuel.

If *any* combination of efficiency gains and renewables succeeds in reducing the rate of burning fossil fuel, then the total CO_2 release will arise mainly from burning in the near term, while it is still large, not in the long term, when it has tailed off. Thus near-term events, which are easiest to influence and foresee, can have the most decisive influence on the CO_2 problem.

CHAPTER 5. BY WHAT POLICY INSTRUMENTS?
5.1. Strategies for implementation: conflicting philosophies

Both economic and political logic favor a strategy of efficiency and renewables. Perhaps because of the diversity of the technologies, uses, and users, recent experience appears to favor a tactic of individual choice and market competition rather than of central planning and mandate. But whichever philosophy of energy policy prevails, correct price signals and the systematic removal of market imperfections will be critical for implementation.

5.2. Price signals

Efficient investment is unlikely unless prices correctly signal abundance or scarcity, free of distortion by subsidy or promotional tariffs. Incremental consumption should as nearly as possible pay its incremental costs. Energy supply industries should finance efficiency improvements and renewables, via loans to their customers, whenever those alternatives cost less than a proposed conventional supply invesment. Properly done, this could simultaneously eliminate the capital burden on consumers, risky overinvestment by the supply industries, and the need to attain very high energy prices before capital can be allocated to "best-buy" energy investments.

5.3. Market imperfections

To ensure that the incentive of correct price signals is matched by opportunity to respond to them, governments wishing to save money, oil, or climatic trouble should identify and purge those institutional barriers—remnants of the cheap-oil era—which prevent people from choosing least-cost options. These barriers include inequitable access to capital and information, certain obsolete and restrictive regulations, and split incentives. Solutions are available but must be matched to local conditions. These problems are awkward, but less so than the alternative set of problems incurred by an economically inefficient policy. Analogous but special considerations apply to implementation in developing countries.

5.4. Accelerating retrofit or turnover of capital stocks

International precedents offer ideas for turning over or fixing up major capital stocks in a few years. The energy and economic benefits of accelerated building retrofits or replacement of inefficient cars are enormous: the US, for example, just by pursuing these two measures in the 1980s, could eliminate its oil imports before a new synfuel or power plant ordered now could even be built, and at about a tenth of its cost. Such programs also offer a policy option for rapidly reducing CO_2 emissions.

CHAPTER 6. HOW QUICKLY CAN THESE THINGS BE DONE?

Deployment rates depend on both political forces, which are conjectural, and technical/economic forces, which appear to give efficiency improvements and soft technologies an advantage of speed over their rivals.

6.1. Rate and magnitude problems

Even ambitious programs of building conventional energy facilities such as power plants and synfuel plants are inherently so slow that, even in principle, they are bound to be too little, too late, to meet traditional projections of demand growth. But efficiency improvements, despite being individually much smaller, have lately been by far the fastest-growing part of energy supplies. The EEC in 1973–78 fueled 95% of its economic growth by energy savings, only 5% with all net new energy supplies. The savings outpaced the supply increases by nineteen to one, delivering more than ten times as much energy as the EEC's increment of nuclear capacity in the same period. Japan has had seven years of GNP growth averaging 4% a year with virtually zero energy growth. In the US, energy savings outweighed new energy supplies in 1979 by better than fifty to one, and in 1980, total energy use fell by about 3½% without reducing GNP. Millions of individual actions in the marketplace—people seeking to save energy to save money—are adding up to massive savings, just as similar individual decisions used to add up to demand growth when real prices were falling.

6.2. Rate constraints and comparisons

The scope for cost-effective energy savings is not only a matter of academic interest; it is a matter of survival for the energy supply industries, which can go bankrupt if they build plants whose output people are unexpectedly unwilling to buy. As investors perceive this risk in new major supply investments, capital is increasingly shifting toward lower-risk, higher-return investments in efficiency improvements. This may in turn constrain supply. But as forecasts of primary demand drop, a given increment of power plants can seldom provide a proportionately larger fraction of the reduced demand, but rather usually a smaller one, because such reductions tend to fall especially heavily on the prospects for electricity sales.

6.3. Comparative rates: theory and practice

Efficiency improvements and soft technologies should be faster to build—in terms of new energy supplied per year or per unit of investment—than "hard technologies," for three reasons: they take much less time to build, can diffuse rapidly into a vast consumer market rather than requiring slow delivery to a narrow utility market, and are held back by institutional barriers largely independent of each other. They can therefore add up by strength of numbers to very rapid total growth, rather than being held back everywhere at once by the generic problems of siting and financing big plants. These theoretical advantages are being borne out in practice by the surprisingly rapid deployment of efficiency improvements in many countries and of renewables in some, notably the US and Japan.

6.4. Oil displacement and other short-term imperatives

Many governments think it is more urgent to replace imported oil now, even with CO_2-releasing coal, than to worry about climate after 2000. But this is a false dilemma. *The same policies can save oil and the climate.* Exhaustive international assessments now suggest that oil has priced itself out of the market: other than new power plants and synfuel plants, it is the costliest competitor, so if all options can compete freely, those uncompetitive options will quickly dwindle, and with them the climatic risk.

If governments wish to speed up the replacement of imported oil, they should choose the package of measures that will do so fastest and cheapest. Investment in very slow and costly measures, such as new power plants, *slow down* oil replacement by diverting resources from more effective measures. The largest, cheapest, and fastest oil displacements will generally be found in its main roles—supplying heat and mobility—rather than in making electricity, which uses only a tenth of all oil (6% in the FRG in 1978). The short lead times of soft-path investments let them start replacing oil—and reducing CO_2 releases—now, not in ten years. Whether they can do this faster than "hard technologies" is independent of whether a country has indigenous fossil fuels, which ones, or how much of them. In all countries so far studied, the fastest and cheapest energy sources are efficiency improvements; then soft technologies; then synthetic fuel plants; and last—slowest and costliest—power stations.

CHAPTER 7. IMPLICATIONS FOR CLIMATIC RISK
7.1. The cost of failure

The risk of climatic change is the largest risk presented by the entire energy system (except for nuclear war). But the foregoing analysis has shown that even if that risk were worth nothing, the same energy policy that minimizes it would still be worth pursuing on other grounds, including lowest economic cost. This rationality does not guarantee its implementation in the face of institutional inertias and powerful forces protecting their own preferred solutions: such choices are not always (some would say they are not ever) based on economic rationality. But however the decision is made, it is not necessary or possible to pursue *all* available options. Different options compete for resources; they build up habits, perceptions, and institutions that inhibit other options; ultimately they exclude each other. If today's remaining stocks of relatively cheap fossil fuels, and relatively cheap money made from those fuels, are not used to capitalize a prompt transition to an efficient and sustainable energy system, that transition may soon become prohibitively difficult. It is essential to make the transition to a climatically benign renewable energy system a rapid one, not only because the fossil fuels needed to get there are becoming scarce and costly, but also because the longer we continue burning them, the less likely it is that the resulting climate will prove congenial.

Thus the risk that continued rapid burning of fossil fuel may change the climate does not mean merely that we may in the coming decades discover we have invested major resources in vast, ponderous energy systems that can no longer be safely used; it may also mean that by then they can no longer be affordably changed. A low-climatic-risk energy strategy offers for free—indeed at negative cost—an insurance policy against such an unwelcome discovery. This insurance would not prevent all dangerous forms of climatic change, but it would, barring gross mismangement, probably prevent those arising from the energy system and let the climatologists concentrate their talents on understanding and preventing the rest.

7.2. Conclusions

Major transformations of the energy system have historically taken a half-century to complete. Today, major and irreversible changes in the world's climate may be set in motion in a much shorter period. This is both because fossil fuels are being burned at a rapid *rate* and because that rate has been *increasing*. How much CO_2 is released is the product of *how fast* fossil fuel is burned times *how long* that rate of burning continues.

This book shows that it is technically and economically possible, consistent with worldwide achievement of the most generous economic goals, to stabilize and then gradually reduce to about zero the global rate of burning fossil fuels. No other known approach can do either, let alone as quickly as 30–50 years.

If this approach were implemented very slowly and poorly, its effect might be only to hold constant the global rate of burning fossil fuel. That would delay for 50 years an increase in CO_2 to about a fifth above its present level (two-fifths above its pre-industrial level). But a far longer delay would be won if the present rate of burning fossil fuel were actually *reduced* through any combination of efficiency improvements and renewables. If it were reduced, for example, from 8 TW to 1 TW over the next 50 years—a goal well short of what our analysis suggests is economically worthwhile, technically feasible, and socially plausible—then that CO_2 level will not be reached for 185 years. Our choice of 1 TW here is not important: if we were not quite so successful—if, say, we reduced the rate of burning fossil fuel in 2025 not to 1 TW but to 2 TW—then that extra 1 TW would raise the CO_2 level in 2025 only by the amount by which it is currently increasing every three years.

Thus the *fact* of the reduction, not its exact pattern or mechanism, is the key to buying time and reducing risk. The combined benefits of a *slower* addition of CO_2 and of its accumulation from a smaller *base value* can then postpone substantial increases in the CO_2 level more or less indefinitely. Whether complete replacement of fossil fuels by renewables is achieved, or when, is from this point of view almost inconsequential, and not a fruitful subject for argument. This is because, at a fossil-fuel burn rate of, say, 1 TW, reaching a CO_2 level twice the preindustrial concentration would take until after the year 3000, when the world will doubtless be quite different anyhow. The delays achievable by this sort of reduction offer almost infinite leisure for squeezing out the last bits of fossil-fuel use. The choice of what combination of efficiency and renewables to use to do this may be facilitated by our review of both.

Reducing the fossil-fuel burn to 1 TW by 2030 is indeed plausible, as is shown by an illustrative global calculation of demand and renewable supply. Such a scenario also shows that without using either Middle Eastern petroleum or nuclear power, an extremely affluent world with doubled population can be fueled in a way which holds the CO_2 level in 2030 to within about 10% of its present value, and which reduces its rate of increase to an eighth of the present rate, dwindling thereafter to zero.

In summary: a highly prosperous world of 8 billion people (or more) will, if it uses energy in a way that saves money, use somewhat less energy than today. That energy can be more cheaply derived from available renewable sources, which are climatically quite benign, than from nonrenewable sources, which are not. Over a half-century or so, starting now, the combined use of efficiency improvements and renewable sources can largely or wholly eliminate the global use of fossil fuels. Even partial achievement of this goal would probably delay the onset of serious energy/climate interactions from its currently expected timescale of a few decades to the best part of a century or more. This stretching out of the time-scale of climatic risk is not sensitive to the exact *pattern* of reducing the rate of burning fossil fuel, providing it *is* fairly steadily reduced, starting soon. The climatic benefits are especially insensitive to whether or when the use of fossil fuel is completely eliminated, providing that it has been reduced to a relatively low long-term level. And finally, all these measures to displace fossil fuel make sense on other grounds, especially economics, and should therefore be done anyway. (In fact, in many countries these technologies are starting to implement themselves much faster than expected.) Since an energy policy with low climatic risks (and low risks and costs of many other kinds too) is *possible, pragmatically achievable, and economically preferable,* governments and citizens can now seek to inform themselves in order to decide whether they would prefer such a policy, and, if they do, to design its details for their own countries and communities.

1
Introduction

1.1. Energy/climate risks: a future problem whose time has passed?

The sun radiates energy at a rate of about 3.9×10^{26} watts (W) (Sørensen 1979a:21,64,173–77). Of this, some half a billionth, or 172,500 TW (172,500 $\times$ 10^{12} W), falls on the top of the earth's atmosphere, and about 81,100 TW reaches the ground. The world's hydrologic cycle is driven by an energy of about 41,400 TW, and the total energy of wind (1200 TW), waves, and ocean currents is several thousand TW. Solar energy is fixed in plants by photosynthesis at a net rate of about 133 TW. The total flux of geothermal heat to the earth's surface is about 32 TW. In contrast, the world's 4.3 billion people directly use energy—not counting the indirect use of solar energy embodied in food and fiber—at a total rate of about 9 TW.[1] This is a great deal of energy in human terms, equivalent to the energy that would be consumed as food by an average of 15 full-time slaves (each eating 3000 kcal/day) for each person on earth; yet it is only a ten-thousandth of the rate at which solar energy reaches the earth's surface. Because of the way in which humankind converts energy, however, that 9 TW is rapidly becoming a significant force in the workings of global climate.

In particular, approximately 8 TW of the roughly 9 TW is

---

1. This figure is uncertain by perhaps 10% because official statistics seldom show the use of noncommercial fuels: a billion people, mainly in developing countries, cook with dung, and 1.5 billion with wood, whose energy yield is at least a substantial fraction of a TW [IEA 1978; Eckholm 1976].

derived from burning fossil fuels—solar energy stored millions of years ago when a tiny fraction of the total plant matter was trapped by freak conditions in an anaerobic swamp where it was protected from oxidation. Through unique conditions extending over geologic time, pockets of fossil fuels totalling over 70,000 TW-y were trapped beneath the earth's surface.[2] It is this patrimony, and especially the mere 720 TW-y of conventional oil and gas resources (including deposits expected but not yet discovered) and the 640 TW-y of coal now considered technically and economically recoverable (Bach 1980a:797), that is currently being burned at a global rate of about 8 TW-y/y, releasing to the air carbon that has not been airborne for many millions of years. Also released are aerosols, oxides of nitrogen, other foreign substances, and heat.

These releases perturb the delicately balanced engine of global climate. Carbon dioxide, in particular, strongly absorbs infrared radiation at wavelengths to which the atmosphere would otherwise be fairly transparent. This extra absorption alters the heat balance within different layers of the atmosphere, and between the atmosphere and the earth's surface. This effect is expected by most climatologists to be the largest and least manageable of all potential climatic impacts of energy use (Bach, Pankrath, & Williams 1980; Bach 1980, 1981), although some localized effects (for example, from spilling oil in the Beaufort Sea [Campbell & Martin 1973]) may also have widespread consequences.

The possible effects of carbon dioxide (CO_2) releases have become a focus for global concern at a scientific and, increasingly, at a policy level, especially in the Federal Republic of Germany and in the United States (Bach et al. 1980; Bach & Breuer 1980; Bach 1979, 1980). An American interagency survey of global environmental and resource problems concluded:

Scientific opinion differs on the possible consequences [of rising CO_2 concentrations], but a widely held view is that highly disruptive

---

2. The 100-odd TW of net photosynthesis today takes place "on current account" and does not alter the earth's carbon balance: whatever carbon is temporarily stored in the standing biomass is soon released as the plant rots or is eaten by a respiring organism, and the carbon, returned to the air as carbon dioxide gas, then becomes available again for new photosynthesis.

effects could occur before the middle of the twenty-first century. The CO_2 content of the world's atmosphere has increased about 15 percent in the last century and by 2000 is expected to be nearly a third higher than preindustrial levels. If the projected rates of increase in fossil fuel combustion (about 2 percent per year) were to continue, a doubling of the CO_2 content of the atmosphere could be expected after the middle of the next century; and if deforestation substantially reduces tropical forests (as projected), a doubling of atmospheric CO_2 could occur sooner. The result could be significant alterations of precipitation patterns around the world, and a 2°–3°C rise in temperatures in the middle latitudes of the earth. Agriculture and other human endeavors would have great difficulty in adapting to such large, rapid changes in climate. Even a 1°C increase in average global temperatures would make the earth's climate warmer than it has been any time in the last 1,000 years.

A carbon dioxide-induced temperature rise is expected to be 3 or 4 times greater at the poles than in the middle latitudes. An increase of 5°–10°C in polar temperature could eventually lead to the melting of the Greenland and Antarctic ice caps and a gradual rise in sea level, forcing abandonment of many coastal cities. (President's Council on Environmental Quality and Department of State 1980:36–37)

Likewise, Working Group I of a 1980 Münster international conference on energy/climate interactions concluded:

Provided energy growth continues at the current rate and with the present mix of sources, there is a definite prospect of a significant climatic change. Along with the expected overall warming of the lower atmosphere there will be shifts in the atmospheric circulation patterns resulting in regional and seasonal temperature and precipitation anomalies. The major effect will not be the creation of a new type of climate, but rather a different distribution of climate. This can have an impact on energy requirements, food production, fisheries, water resources, forestry, land use, transportation, tourism, and human health and well-being. In short, the total global socioeconomic structure is vulnerable in varying degrees to . . . climate change and variability. (Bach, Pankrath, & Williams 1980:xiii)

Although climatic change is but one of a long list of pressing resource and environmental issues (President's Council on Environmental Quality and Department of State 1981), it commands attention by its scale, pervasive consequences, and virtual irreversibility. "The known fossil fuel resources are large enough to result in peak CO_2 levels 4 to 8 times the pre-industrial value within 2 to 3 centuries—assuming, of course, increased usage of these fuels. Such high levels will only slowly decline so that a CO_2 concentration perhaps twice the pre-industrial is likely to persist for over a thousand years." (Bach 1980b:3) Since the same combustion processes that produce CO_2 also produce much of the other pollution (by trace gases and aerosols) that would tend to reinforce the warming effect of the CO_2 (ibid.), the CO_2 problem can be taken as a surrogate for the entire set of energy/climate problems. Though not a comprehensive view, this is near enough to be a useful beginning.

The CO_2 problem is especially dangerous because the relevant climatic mechanisms involve considerable time-lags—temperature increases alone are expected to lag behind the CO_2 increases that drive them by about 10–20 years (Bach, Pankrath, & Williams 1980:xv)—and because the changes expected may be incremental or subtle enough to be difficult, without accumulating a long time-series of records, to distinguish from normal fluctuations. There could then be even longer delays in obtaining domestic and international recognition of the problem, agreement on remedies, and effective implementation. Combined with the inherent momentum of the energy system, where major changes have typically taken 50 years, these delays risk encountering major climatic changes only after it is too late to avoid them. The pervasive scientific and social uncertainties about climatic change also argue for prudence and restraint in energy actions that might increase climatic risk. Thus the Münster conferees recommended a "low-climate-risk energy policy, which would promote the more efficient end use of energy, secure the expeditious development of energy sources that add little or no CO_2 to the atmosphere, and keep global fossil fuel use, and hence [the rate of] CO_2 emission, at the present level." (ibid.:xix) But this approach has not yet noticeably influenced CO_2 policy.

The consequences of a near-doubling, let alone a larger increase, of the level of CO_2 in the atmosphere are probably too large to be safely determined by experiment (ibid.:xx–xxiv;

President's Council on Environmental Quality 1981; Kellogg & Schware 1981; Schneider 1980, 1981; Schneider & Chen 1980). Accordingly, considerable scientific talent and resources are being devoted in several countries to trying to *predict* the climatic effects of various CO_2 levels and, in turn, the likely effects of those climatic changes on various human activities (U.S. Department of Energy 1980a,b,c; Bach, Pankrath, & Williams 1980; Bach et al. 1980; Kellogg & Schware 1981). But at best a few percent—probably less than one percent—of this effort is devoted (U.S. Department of Energy 1980) to analyzing *how much fossil fuel is likely to be burned* in the future—the crucial data on which all other results must rest.

Perhaps there seemed no need to consider this question in detail, because in the mid-1970s most official forecasting bodies agreed that the rate of burning fossil fuel would rise rapidly, continuously, and indefinitely. Because all the forecasts rose so steeply, minor differences between them meant only that the CO_2 level would double a few years earlier or later. Climatologists took this virtual consensus to indicate that the forecasts were correct, assumed therefrom that CO_2 was going to be a serious problem sooner or later, and worked hard to determine when it would become serious. By about 1980 they had largely agreed, as the previously quoted Münster Working Group I report concluded, that the problem was already serious enough that governments should promptly seek to flatten out the growth in the rate of burning fossil fuel.

Meanwhile, however, from another discipline—the engineering/economic analysis of energy efficiency—word was coming in that the energy demand forecasts on which this elaborate superstructure of climatological analysis had been built were no longer defensible, because the trends they embodied were no longer in society's economic interest. All the new indications are that our economic interests would be best served by a "least-cost energy strategy" (Sant 1979, 1981) which emphasizes dramatic efficiency improvements and, secondarily, the rapid deployment of appropriate renewable sources. This news, expressed in careful and detailed studies from more than a dozen countries, is still reverberating in the energy policy community and has not yet reached most climatologists. Yet it implies that the problem they have been so assiduously studying probably need not arise: the ever-expanding fossil-fuel burn that produces the problem could result only from an economically inefficient energy policy that seems increasingly implausible and at variance with rapidly emerging trends.

When one integrates these new energy policy conclusions with climatological data, one sees that an economically efficient energy policy buys disproportionate time—so much that if harmful effects from CO_2 are not already virtually upon us (Hansen et al. 1981), they can probably be postponed for centuries or avoided forever. "The major bonus of such a low risk policy is that in the best case it may prevent climatic impacts altogether, and that in the worst case valuable time is gained to obtain better information to redirect policy [and in which to achieve that redirection]. The good message of this pragmatic policy gives rise to modest optimism, since it is based on measures that make sense also for other than climatic reasons and should therefore be taken anyway." (Bach 1980b:5).

Conversely, *not* following such a policy would be all but certain to lead to serious climatic impacts too soon to avoid them except by prompt, concerted policy shifts on a global scale. Continuing outmoded energy policies incurs the risk that "energy decisions based on . . . poor information may lock society into energy paths . . . devoid of any flexibility once energy systems with long market penetration times [such as synthetic fuels] have been adopted." (ibid.: 5)

Whether the CO_2 problem *is* urgent, then, depends on something not yet properly assessed: how much fossil fuel will be burned. Even climatological research as valid and valuable as much that has been done in the past few years cannot be useful for policy unless its energy data are up-to-date. That is the purpose of this book. As its precursor remarked, until the mid-1970s, energy demand was forecast largely by the mechanical application of a straightedge to semilogarithmic graph paper—a method often replaced today

> . . . by teams of systems analysts armed with computer models whose elaborate opacity generally conceals primitive methodology and

unrealistic assumptions. These extrapolative models tend to predict large and rapidly rising demands for energy, especially in premium forms, hence a need to build huge new energy supply systems with increasingly nasty problems of many kinds, including risks of climatic change.

Before devoting much effort to studying these risks, we should carefully explore whether we need to incur them. How much energy *will* we need? In what forms? At what scales? and in consequence, Where from? How fast? Serious attention to these prior fundamentals reveals a very different result (and a realm of what Pogo called "insurmountable opportunities").

Greatly increased affluence, which this analysis assumes for all countries, does not require proportionate increases in energy use, because energy productivity—the amount of work wrung from each unit of energy—can increase even faster, at far lower cost than new energy supply. The next best buy is generally the appropriate renewable energy sources, which appear ample for virtually all long-term needs without any technical breakthroughs. Despite increasing wealth, then, the worldwide rate of burning fossil fuels can stay roughly constant in the short term and decline gradually, over about 50–75 years, to about zero, thus minimizing climatic risks of burning those fuels.

Such policies are both simpler to understand and technically more sophisticated than their predecessors. They can be relatively free of climatic risks in themselves, and at the same time offer faster displacement of oil and gas, longer-term sustainability, lower economic and political costs, lower risks of technical failure and of the proliferation of nuclear weapons, more tractable safety and environmental problems, and more favorable social side-effects than traditional, less economically efficient policies. These conclusions depend very little on geography, climate, culture, political system, level of economic activity or technical development, and density of population or industry. (Lovins 1980:1–2)

This book will explore the foregoing thesis in greater detail, showing how the world rate of burning fossil fuel can be not merely stabilized but reduced, and how such a reduction can be not inimical to, but rather a principal engine of, economic growth. We shall also estimate the effects on CO_2 releases and hence on climatic change, both 50 years hence and asymptotically, of economically efficient investments to meet global energy needs.

1.2. How much time can we buy?

The disproportionate time-buying benefits of burning fossil fuel more slowly can be illustrated in a slightly more technical fashion by calculating how soon, given various trajectories for the fossil-fuel burn, atmospheric CO_2 concentrations would attain certain levels. To do this, we must make several simplifying assumptions[3] whose net effect is probably to make CO_2 problems look somewhat less urgent than they actually are. These yield the following indicative projections (Table 1.1) for the dates of attaining various CO_2 levels, shown as ppmv (parts per million by volume) and as multiples of the 1981 level—about 338 ppmv—and of the estimated preindustrial (say 1850) level—about 280 ppmv.

The pattern of dates for this array of fossil-fuel burn trajectories—exponentially rising, constant, or linearly falling to various levels by 2025—is remarkably revealing (Figure 1.1). Even though nobody knows which of the nominal CO_2 levels shown (if any) is really of climatic concern, it is still possible from such an illustrative calculation to infer valuable lessons for policy.

First, the vertical column under "350 ppmv" shows that if a CO_2 level of about that magnitude is of climatic concern, it is too late already. Short of an immediate, worldwide crash program of conservation, solar energy development and reforestation, aimed mainly at the major energy-using countries, one could only slightly mitigate the climatic effects. Since the lead time for building major fuel-converting and fuel-using facilities is generally longer than the dates in this column, major investments might also have to be written off.

Conversely, the horizontal row assuming a 4%/y exponential rise shows that such rapid growth is a recipe for events' outrun-

---

3. We assume for this exercise no CO_2 source other than burning fossil fuel; 55% of released CO_2 stays airborne; this fraction remains constant (i.e., no de- or reforestation or changes in atmosphere/ocean exchange); 1 Pg (10^{15} g) of carbon released to the atmosphere, or 0.55 Pg remaining there, increases the global average CO_2 concentration by 0.26 ppmv; and each TW-y of heat released by burning fossil fuel releases $(0.605 + 0.0017/y)$ Pg of carbon. This linear form approximately fits a plausible 50-y shift in fuel mix, for example as shown in Table 1.2.

Table 1.1 *Carbon dioxide levels as a function of fossil-fuel burn trajectory*

Assumed Global Fossil-fuel Combustion				Dates of Attaining CO_2 Levels					
1975–2025 pattern of increase or decrease (1975 ≡ 8 TW)	rate (TW) in 2000	rate (TW) in 2025	total TW-y 1975–2025	ppmv=	350 1981× 1.04 1850× 1.25	400 1.18 1.43	450 1.33 1.61	500 1.48 1.79	550 1.63 1.96
+4%/y compounded yearly	21	57	1270[a]		1986	2003	2013	2020	2025
+3%/y compounded yearly	17	35	929		1987	2006	2018	2027	2033
+2%/y compounded yearly	13	22	690		1987	2010	2025	2037	2046
+1%/y compounded yearly	10	13	521		1988	2016	2037	2053	2067
constant 1975 level	8	8	400		1989	2025	2059	2090	2119
constant to 1985, then linear @ -0.1750 TW/y	5.4	1	260		1989	2166[b]	2444[b]	2723[b]	3001[b]
same @ -0.1975 TW/y	5.0	0.1	242		1989	3606[b]	6389[b]	9172[b]	
same @ -0.2000 TW/y	5	0	240		1989	never[c]			

a. Exhausts all estimated ultimately recoverable oil and gas, plus nearly 90% of coal now considered technically and economically recoverable (Bach 1980a:797).
b. Assuming that the 2025 burn rate and release coefficient (Table 1.2) remain constant at the 2025 level after 2025.
c. Converges asymptotically to about 370 ppmv.

Table 1.2 *Possible future shifts in fossil-fuel mix and CO_2 coefficient*

fuel	% share in: 1975[a]	2000	2025	PgC emitted/TW-y[b]
oil	46	30	15	0.60
gas	24	20	20	0.43
direct coal	29	45	55	0.75
coal/shale synfuel	<1	05	10	1.02
composite PgC/TW-y	0.605	0.655	0.691	(weighted averages)

a. JASON data from Elliott & Machta (Committee on Governmental Affairs 1979:72); includes the flaring of natural gas.
b. MacDonald data from Woodwell et al. (ibid.:116).

The short-term shifts shown in the mix are consistent with official energy policies; longer-term shifts, with depletion curves reasonable for high demand. With low demand, as noted in Chapter 7, the coefficient need not increase.

Figure 1.1 *Approximate projections of atmospheric carbon dioxide levels as a function of the pattern of increase or decrease in the rate of burning fossil fuels, 1975–2025*

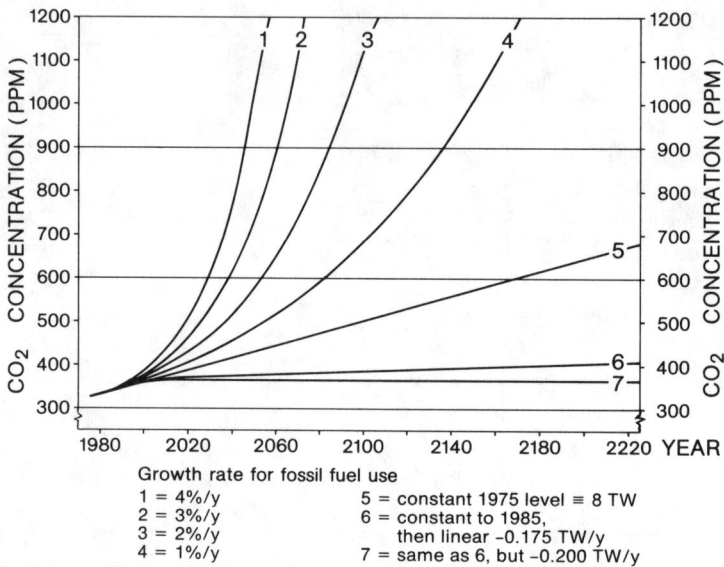

ning our ability to manage them—to say nothing of providing the fuel itself. (The first footnote to Table 1.1 shows how such growth would strain the resource base. There is perhaps 16 times as much total coal resource in the ground as is now considered recoverable, but if it were all recovered and burned, and if half the resulting CO_2 stayed airborne, global CO_2 levels would increase approximately sevenfold. Hence long-term global coal use is almost certainly limited by climatic effects, not by the coal resource itself.)

The most useful lessons from Table 1.1, however, appear if we consider that ordering, say, a coal-burning power plant or synfuel plant today commits us to use coal until about 50 years from now, and that it will probably also take about 50 years to replace essentially all the fossil fuel with climatically benign energy sources. If, therefore, we take 50 years as an indicative planning horizon, Table 1.1 suggests these further conclusions:

1. The rate of burning fossil fuel must not increase more than

4%/y if the CO_2 level that causes major climatic change is 550 ppmv; nor about 3.3%/y if it is 500 ppmv; nor 2%/y if 450; nor 0%/y if 400. That is, as we consider more proximate CO_2 "alarm thresholds," the tolerable rate of growth of the fossil-fuel burn drops *nonlinearly, more and more steeply* (Figure 1.1).

2. Since it is not now possible, may not soon be possible, and may never become possible to predict with confidence which (if any) of these nominal levels is a realistic "alarm threshold," *a constant rate of fossil-fuel burning is the highest tolerable rate* if we wish to keep safely beyond our minimum planning horizon the lowest nominal CO_2 level whose attainment is not already virtually unavoidable. Of this option, Bach (1980b:3) points out: "If . . . [the rate of] fossil fuel consumption could be kept at the present level, a 50% CO_2 increase, corresponding to an average temperature increase of 1–1.5 [C°], would still result in AD 2100. Such a gradual temperature increase over such a long time period [as opposed to about 3C° by the mid-2000s on conventional projections] is perhaps tolerable. If not, at least time is bought for taking remedial action."

3. Even at an "alarm threshold" of 400 ppmv, a comfortable *several centuries* of time for a supply transition can be bought (as Figure 1.1 illustrates) by *reducing* the rate of burning fossil fuel, over about 50 years, to about 0–1 TW. If this were begun during approximately the 1980s, the bulk of the savings in CO_2 "release commitment" would have been made by about the year 2000, and the exact details of how low the fossil-fuel burn would eventually go, and how long it would take to get there, would be relatively unimportant. (As fossil-fuel use tailed off, the remaining burn would become rapidly less consequential.)

4. Such stabilization and reduction of the fossil-fuel burn can be won by *any combination* of efficiency improvements and alternative sources. The latter are generally slower and costlier to build than the former, but can benefit from the longer time available to build them if some fraction of the fossil-fuel reduction has first been done by efficiency gains. (Algebraically, if that fraction is x, the time available for renewable deployment increases by a factor of $1/[1-x]$.)

5. The *disproportionate time bought* (Figure 1.1) by reduc-

tions in the rate of burning fossil fuel suggests that a strategy of efficiency improvements and renewable sources—shown below to be the cheapest and fastest method—deserves a high priority for reducing climatic risks in the face of uncertainty.

In short, although conventional analyses of climatic risk *assume* rapid and continuing growth in the global use of fossil fuel, we shall show that the rationale behind this assumption cannot withstand rigorous scrutiny. More careful analysis of the amounts, types, and sources of energy that can provide desired energy services at least economic cost leads to a robust conclusion: economically efficient energy investments will at least hold constant, and probably reduce, the world rate of burning fossil fuel, starting essentially immediately, even assuming rapid worldwide economic growth and complete industrialization of the developing countries. Table 1.1 implies that if this result is achieved, then even a 20% increase in CO_2 level from its present value will probably not occur for at least 50 years or so, and may well be postponed for centuries or forever. The otherwise inexorable increase in the rate of burning fossil fuel would be stopped virtually in its tracks.

We next explore the fast-moving developments in energy technologies and economics that make possible such an encouraging result.

2
How Much Energy Will We Need?

"It is difficult," says Niels Bohr, "to make predictions, especially about the future." This difficulty has never inhibited those eager to forecast the energy future: many forecasts embody such spurious precision that, as Meghnad Desai once astutely remarked, we appear to know much more about the future than we do about the present or the past. Analysts concerned with precision in 2030 must first struggle with the reality that even in the best-ordered countries, basic energy data for 1975 may disagree by tens of percentage points. Any effort to look into the energy future must start, then, with a certain humility.

But this does not imply that nothing meaningful can be said about energy needs more than a few years ahead. On the contrary, careful application of engineering economics can yield forecasts for decades ahead that are more firmly based in reality than most extrapolative attempts to forecast markets and prices next year. *Purely* economic analysis of past behavior cannot reveal the full range of technical options available, because both technical developments and the market imperfections that inhibit their use may change. Engineering economics, however, shows what is technically possible and economically worthwhile to do. A comparison of these possibilities with past performance under specified market imperfections can then give a quite percipient impression of how quickly people are likely to take advantage of newly available choices. Without presuming to forecast what *will* happen, we can therefore say with some precision what *could* happen if particular policies were implemented to provide efficient marketplace incentives and opportunities.

The future is more uncertain that it used to be. Unforeseen technical and political developments occur ever more frequently. To cope with these uncertainties, our assumptions in this analysis embody many "conservatisms"—deliberate biases in the directions least favorable to our conclusions. More realistic assumptions would therefore reinforce those conclusions, not vitiate them. To ensure that our calculations can be understood, verified, and (if the reader desires) modified to test their sensitivity to key assumptions, we shall document our assumptions from empirical measurements of cost and performance, and seek to make the calculations as "transparent" as possible. On the other hand, to avoid drowning the reader in literature, we shall select only those sources necessary to make the point, not the enormous body of supporting measurements. Readers desiring access to the full range of international literature can enter it via the extensive citations of our sources, and can keep up to date via the international bimonthly periodical *Soft Energy Notes*. We present here an explicit, policy-oriented, descriptive (not prescriptive) exploration of a pragmatic energy future—seeking to fill what John Steinhart called "the enormous gulf between the unavoidable and the miraculous."

This chapter explores economically efficient energy futures on a national, regional, and global scale over the next half-century or more—the time-scale over which major energy facilities ordered today will operate. It deals only with the need for energy in various forms, not with the sources used to meet those needs: subsequent chapters will evaluate energy supply and estimate the combined effect of energy efficiency and non-fossil-fuel sources on total CO_2 releases. Our aim, to repeat, is not to *predict* the future but to offer a practical alternative to economically *in*efficient futures which, if no other option were available, might seem inevitable. It is those high-energy futures which have so concerned scientists aware of the fragility of global climatic balance.

Earlier discussions of energy/climate interactions rested on forecasts of long-term world energy needs which were, if not sophisticated, at least plain: that in 50 years or so, the rate of human energy use will have risen exponentially (or by a gradually saturating logistic curve) to 50–80 TW, about six to ten times the present rate. Recent major studies of demand in 2030 have ranged from a low of about 16 TW (Colombo & Bernadini 1980), a doubling of the present rate, to a mid-range of 22 TW (Rogner & Sassin 1980; IIASA 1981), the IIASA "low scenario," to a high of 36 TW (ibid.), the IIASA "high scenario." Our analysis, expanding arguments by Lovins (1980), will show that on the contrary, based on the best available technical evidence about cost-effective improvements in the efficiency of using energy, a doubled world population—8 billion people—can attain the same degree of material well-being as assumed in the IIASA scenarios but using only 8 TW, and perhaps, in a century or so, less than 4 TW. Later chapters will show that these figures imply no fossil-fuel burn.

2.1. Defects of conventional forecasts

Before we present that analysis, however, it may be useful to provide some perspective by reviewing the main reasons for the two- to tenfold discrepancy between our demand estimates and some others. The higher forecasts, on which governments and international organizations have so far based their expansionist energy policies and their consequent concerns about climatic risk, arise from an almost universal failure to account properly for the following twelve effects:

1. *Price elasticity of demand* (i.e., the more one charges people, the less they buy). Most analyses ignore or minimize long-term adaptation to higher price. The IIASA projections (1981), for example, have very low implicit price elasticity of demand: that is, they assume that the very high energy prices needed to pay for the projected supply system will not lead energy users to substitute brains, capital, and other resources for much of their energy. This internal economic inconsistency—assuming demand so high that meeting it requires a supply system so costly that its energy output would be expensive enough to elicit far greater gains in energy productivity—is unfortunately very common. IIASA, however, carries it to the

unusual length of apparently assuming, in some cases, a *positive* implicit price elasticity—that is, higher price will *increase* demand (Lovins 1981).[1] IIASA assumes energy/GDP elasticity to be substantially higher (worse) than most OECD countries have actually achieved in recent years, and to *rise* in North America between 1985 and 2015. Despite rising prices, the technical efficiency of some sectors appears to decrease *below* its present level.

2. *Inverse price elasticity of demand.* Many studies project from past trends in energy demand in periods when real energy prices fell, as gasoline prices did in the UK (by 14%) and the Netherlands (by 7%) from 1970 to 1979, or as average US residential electricity prices did (by about 80%) from 1940 to 1970. IIASA goes further still, assuming that energy/GDP elasticity will improve at a similar or slower rate in the future, with rising real prices, than it did in the past when real prices declined (ibid.).

3. *Subsidies.* Nearly all studies assume continued massive (and unevenly distributed) subsidies to make energy, especially additional energy, look cheaper than it really is to the society. The size of the historic and current subsidies via both tax and pricing policy is unknown except in the US, where subsidies on current account total over $100 billion per year (Lovins 1981a)—enough to reduce average energy price by over a third and nuclear electricity price by over half—and historic subsidies (Cone et al. 1980) total over $252 billion (in 1978 dollars), with

---

1. The IIASA model's price elasticity is "implicit"—derivable after the fact but not fed into the model explicitly. (The "model" is actually a combination of several disparate and incompatible models which can be used in sequence only by having an operator subjectively adjust the output of one to match the needs of the next and compensate for the deficiencies of each. These adjustments largely determine the output: hence the whole arrangement is more accurately described as a simulation game. It is in any event what is called in the trade a "goulash model," whose differing parts—econometric, input-output, linear-programming, and accounting submodels—cannot be expected when used in sequence to yield any meaningful results [Meadows 1981].) Demand is generated exogenously, not by the internal workings of the model: the operator must provide 36 constants per region and 146 coefficients per scenario-year, and the model merely does the bookkeeping to convert these thousands of externally supplied data into total derived demands for various fuels. The operator is free to use unrealistic or inconsistent assumptions, as was apparently often done (ibid.).

the nuclear term alone sufficing to cut the apparent cost again roughly in half (Bowring 1980). Historic subsidies have also been assessed in Japan (Mitsubishi Research Institute 1978) to be proportionately greater than in the US, totalling some ¥37 trillion (in 1976 yen) net during 1946–76. Similar studies, funded by the US Department of Energy via Battelle Pacific Northwest Laboratory and being done in the Battelle offices in Frankfurt and Geneva for the Federal Republic of Germany and France respectively, are not yet available. Follow-on funding has apparently been terminated.

4. *Saturations* may physically constrain growth in many traditional energy markets. Official UK projections of space-heating demand—the largest component of national energy use—have until recently assumed that heating is an unlimited linear function of income; but the rich may not wish to roast. Car traffic is ultimately limited by one's unwillingness to sit in a car all day and by one's inability to drive more than one car at the same time (though some traffic forecasts, forgetting this, have assumed old-age pensioners driving about for many hours every day). As Basile (1981) remarks, even GDP growth itself raises such awkward questions: IIASA's assumed 85–255% increase in real per capita GDP for North America by 2030 is "hard to imagine" (five cars and a boat and a helicopter in every garage?), although this study, for the sake of argument, tactfully ignores such economic saturations.

5. *Promotion.* In most countries, rapid growth in energy demand, especially in such fuel-intensive sectors as electric-resistance space heating, has been heavily promoted by advertising, concessionary tariffs, cross-subsidies from other classes of users, and so forth. Even where these distortions are now being removed, they markedly affected past behavior, which is accordingly an unreliable basis for extrapolating efficient choices in the future.

6. *Distortions in basic data.* Many studies rely on outdated and excessive estimates of population, labor force, labor productivity, and so on. It is also common to derive future GDP growth assumptions from projections of labor productivity without regard to whether higher labor productivity will increase or

decrease total factor productivity. (That is, it may be possible to achieve more efficient production overall by giving more emphasis to the productivity of capital, energy, or other inputs rather than of labor alone.) Some nations have unique distortions: the US annual energy growth rate in the 1960s was inflated 1–2 percentage points by the Vietnam war, leading to an enormous error if extrapolated for several more decades. Denmark has anomalously high demand for jet fuel because long-haul SAS flights tend to tank up in Copenhagen. Conversely, using Switzerland as a basis for extrapolation to other countries could lead to an underestimate of demand, since the structure of the Swiss economy is such that the net import of "gray" energy embodied in goods and services is about a third as large as the total "black" (direct) energy which shows in the statistics. Finally, even in industrialized countries, noncommercial or unconventional fuels often make a substantial but unrecorded contribution. The United States in 1980, for example, got about twice as much delivered energy from wood as from nuclear power (Lovins & Lovins 1980:66n144; RTM 1981), but none of the wood appeared in official supply statistics.

7. *Structural changes.* Worldwide changes in the terms of trade may shift production into areas of greater comparative advantage. Such structural shifts were the largest source of reductions in the energy intensity of French industry during 1960–78 (CGP 1981:21). In the Federal Republic of Germany, projected reductions in industrial energy use over the next 50 years from officially anticipated shifts in composition of output are nearly as large as from decreases in intensity per sector, but only part of the latter and none of the former is reflected in the government's energy demand forecasts (Krause et al. 1980).[2] The primary energy used per unit value added in West Germany is

---

2. A related policy problem is that owing to inertias in the redistribution of labor, governments may be tempted to subsidize exports of products in which they enjoy no comparative advantage. In consequence of Lerner's Symmetry Theorem, however, such export subsidies produce no net economic benefit to the exporting country, but only a redistribution from unsubsidized to subsidized *domestic* industries. This is because floating exchange rates seek equilibrium, causing each subsidized export to be counterbalanced by an opposite and roughly equal "import rebound" which cancels any balance-of-payments benefit (Office of Management & Budget 1975).

some 35 times as large in primary metallurgy as in, say, commerce and banking, so broader shifts between secondary, tertiary, and quaternary sectors can also profoundly affect energy needs. Further, the capital contribution to production functions varies enormously—up to fivefold for some processes (Dunkerly et al. 1977)—according to the relative prices of labor, capital, and intermediate inputs such as energy. Further distortions are introduced by treating gross economic activity as if it measured welfare: in fact, it omits important non-market economic activities but includes remedial and transaction costs, so that for all we know, increasing GDP, especially in a highly industrialized country, might even *decrease* net welfare. In any event, to the extent that welfare is derived from a stock of material artifacts, what GDP growth (a common measure of economic performance) actually measures is not the size of that stock but its second time derivative—the rate of change of a *flow* contributing to that stock, and a flow which, moreover, one should arguably be trying to minimize, not maximize, per unit of stock maintained (Daly 1978), in order to achieve the greatest fixed wealth with the least trouble. To avoid such pitfalls, energy intensity must be measured by "technical coefficients"—energy used per *physical* unit of specific, tangible product or activity. To ensure specificity and concreteness, we use this approach in the following analysis.

8. *Primary energy projections* cause increasingly serious errors because, with an assumed shift towards electrification and synthetic fuels, more than half of officially projected growth in primary energy supply in the next few decades (e.g., in West Germany, France, the UK, and the US) will go to conversion and distribution *losses,* masking much slower growth in final consumption or "end-use energy." Forecasting should instead be done in end-use terms and primary energy treated as derivative, depending on the supply systems assumed. This methodological problem is important in the IIASA model (1981), where greater central electrification is structurally assumed rather than objectively derived or economically justified (Lovins 1981; Meadows 1981), and "low"-scenario conversion *losses* in 2030 are nearly as large as *total* world primary energy use today.

9. *Excessive aggregation* conceals crucial engineering details of potential savings. Because energy is used in billions of devices of thousands of types, the efficiency improvements possible for each type are individually small compared with national demand, but can add up rapidly: this is, after all, precisely the way in which a myriad individual demands add up to the national total. Homogenization—not disaggregating demand sufficiently to detect most of the terms—thus ignores cumulatively large savings. Further, as extensive international analysis has shown over the past two years, some important opportunities for meeting energy needs at low cost are so fine-grained that they are simply invisible in a "top-down" analysis, and can become detectable only by working "from the bottom up" in a style highly disaggregated not only by energy use but by locale. Reddy (1980) has strikingly illustrated this in the case of India, where neither official nor alternative energy forecasts seem achievable on a national basis, but where by aggregating from the village scale upwards, new opportunities for integrating energy, food, water, and shelter systems can reveal practical solutions.

10. *Technological progress.* Few forecasters, even in technologically sophisticated institutions, appear to be aware of the extraordinarily rapid progress made in recent years, mainly in the private sector and outside official information channels, in doing more work with less energy. The IIASA study (1981; Basile 1981), for example, considers it a great achievement to have by 2030 in North American a car fleet average of 35 miles per gallon—the same as the average 1981 import sold there, and less than half as good as existing Volkswagen prototypes. Similarly, a 40% improvement in house thermal efficiency is seen as difficult to achieve, even though basic weatherization alone empirically yields a 50–65% gain costing \$1/GJ (Socolow 1978; ACEEE 1981), clever retrofit techniques can often save 90% or more with a payback time of less than ten years, and space-heating demands of virtually zero in new buildings are highly cost-effective today (Lovins & Lovins 1980), as noted in detail below. Lacking such information, the IIASA analysts find (Basile 1981) that the projected increase in North American liquid-fuel and primary energy demand seems "irreducible."

11. *Asymmetrical assumptions*. Analysts familiar with and fond of one class of technologies often tend, perhaps unconsciously, to assume more favorable costs, deployment rates, roles, and performance data for their favorites than for less familiar or less congenial technologies. This sort of bias pervades the IIASA study's assumptions (1981), which tend to use cost and performance data better than empirical data for some technologies and worse than empirical data for others which the analysts like less or know less about. The discrepancy between wish and reality is especially marked—nearly twofold at the margin (Komanoff 1981)—for nuclear reactors, which the IIASA group has apparently assumed *do* cost what they *would* cost if society did not demand the present safety requirements. (This might be called an "unlearning curve" or a "forgetting curve.") In contrast, quite different criteria and level and strength of analysis are applied to appropriate renewable sources, which the IIASA group did not even begin to study seriously until late 1980, yielding preliminary results quite inconsistent with those assumed in the study (Caputo 1980, 1981).

12. *Failure to provide symmetrical comparisons at the margin.* Even with correct data, many analyses fail to test technological options against each other using fair and consistent criteria—the sort of test that would occur in an ideally competitive marketplace. In the US, for example, no official study has done this, and national policy has favored the worst buys first—starting at the top of the supply curve and working down. The fallacy of this approach has been definitively shown by Sant (1979, 1981) and SERI (1981), as described further below. Without such a symmetrical comparison, technologies may also not be given an opportunity to perform properly because of internal inconsistencies in the analysis. Renewable sources, for example, may be assessed high up on the steeply rising portions of their supply curves,[3] where no kind of energy supply makes economic sense. But a cost-effective investment strategy (ibid.) would first do cheaper efficiency improvements—thus bringing the renewables down to the shallowly sloping lower portions of their supply

---

3. These show the unit cost of additional supply as a function of the amount demanded.

curves where they cost less and perform better (owing to synergisms described in Section 4.2.1). The IIASA study (1981), like many others, does not do so.

2.2. Need for improved forecasts

Reliance on projections embodying any or all of the foregoing defects will lead to serious exaggeration of future energy needs and hence of the scope and urgency of climatic risks. (It can also lead to erroneous commitments of vast resources to build energy supply systems which then turn out to be unnecessary but whose costs cannot be fully recovered from the reduced revenues derived from lower-than-expected sales. This can induce the supply industries to raise prices still further, thus depressing demand growth—or even the level of demand—even more, perhaps enough to decrease revenues. The resulting "spiral of impossibility" can prove ruinous to both the supply industries and their customers [Lovins 1981b, c].) Both sound national energy policy and a balanced assessment of climatic risk require more sophisticated, detailed, and disaggregated assessments of future energy needs: analyses which take full account of the best state of the art, pay more attention to physics and engineering detail, and systematically apply rigorous economic tests of cost-effectiveness.

Since 1978, such studies have been done, mainly in the private sector, in a wide range of industrial countries. They

- are notable for their physical concreteness: they construct end-use needs from the bottom up and require the analyst to be explicit about what the energy is expected to be used for;
- require only a hand calculator to compute;
- have the virtues of being simple, transparent, scrutable, and ratproof;
- are economically conservative, basing their decision rules not on political ideologies or thermodynamic idealism but on the orthodox market criterion of minimizing private internal costs;
- assume a high, and (especially for the world's poor) a vastly increased, level of industrialization and material affluence—a level unprecedented in the history of the world—rather than

the agrarian utopia, primitive hardship, or asceticism often ascribed to such analyses by critics who have not read them;
- nonetheless show straightforward, economically attractive ways to lower primary energy/GDP ratio[4] by a *factor of three to six or more* in some of the most advanced countries, at a more or less linear rate over many decades; and
- achieve this entirely through "technical fixes": well-known, currently available technical measures that are now economic (at least against marginal prices) and have no significant effect on lifestyles. Technical fixes use energy more efficiently to provide unchanged or increased services, rather than saving energy by curtailing or foregoing those services. These two ways of saving energy are quite different—insulating your roof does not mean freezing in the dark—although they are often confused.
- Finally, most such studies reflect an unwarranted degree of cornucopianism in dealing with resources other than energy (not to mention the satiability of human needs): by assuming that energy efficiency gains will be used mainly to fuel rapid, traditionally industrialized economic growth, they avoid—just as conventional growth policy does—dealing explicitly with awkward questions of distribution and sustainability. This omission (to which there are notable exceptions) only strengthens the conclusions.

These analyses of what we shall call "efficiency scenarios"—efficient in the use of both energy and money—were not, in general, done with climatic problems in mind, but are readily applicable to them. They range from mere suggestive sketches to perhaps the most precise and persuasive quantitative analyses of the energy future that have been done anywhere (Olivier et al. 1981; Leach et al. 1979; CONAES 1979; Brooks 1981; Nørgård

---

4. We use E/GDP ratios in this paper only as a convenient shorthand for a far more complex set of disaggregated technical coefficients. Such ratios are *not* suitable for rigorous comparisons over time or between countries, for reasons explained elsewhere (Lovins 1978:190–191; Krause et al. 1980:120–121)—notably an inability to distinguish the energy intensity of individual functions from those functions' weighting in the composition of output of each economy.

1979; US Department of Energy 1980d; Krause et al. 1980; Krause 1980; Sant 1979, 1981; SERI 1981). Several dozen scenarios are available (IPSEP 1980, 1981) from more than 15 countries. Yet *even one* such study, for a diverse, heavily industrialized economy in a rigorous climate (such as northern Europe), serves as an existence proof—an *a fortiori* argument—for other countries. We shall now offer such an example.

2.3. Case study for the Federal Republic of Germany

As a basis for exploring the long-term energy requirements for meeting equitably the needs of an expanding world population, especially in the developing countries, we describe here how to reduce roughly fivefold, by cost-effective technical means, today's total energy needs per capita in an economy producing an enviable level of goods and services: the Federal Republic of Germany (FRG). Besides the practical convenience of familiarity, our use of the FRG for this case study offers several theoretical advantages:

- At the date we shall use (for reasons described below) for our base case—1973—the FRG was, among the highly developed nations, the one with the highest fraction of GNP derived from industrial production (50.7%). It thus represents most closely the structure of energy demand which might be experienced by a large number of "threshold economies" as they industrialize further.
- The FRG is already one of the most energy-efficient industrial countries in the world (Schipper 1978; Darmstadter 1978). Improvements feasible in the FRG should therefore be smaller than those feasible in most other countries.
- Most countries have a far less rigorous climate than the FRG. Half of all the delivered energy needs of the FRG today are for heat below 100°C, nearly all for space-heating buildings. The great bulk of the world's people live in climates needing little or no space heating for comfort.[5]

---

5. Nor space cooling, if the buildings use traditional hot-climate designs. The measures described below for saving heat also prevent overheating. Trees, window overhangs, other passive design features, and if needed such special measures as earth-pipe cooling or roof

- Input-output analyses have shown (Bauerschmidt 1978) that the FRG is a slight net exporter (surplus ca. 5%) of energy embodied in goods and services. Thus present energy use statistics are not significantly distorted by trade.
- In the FRG, the prospect of declining population and the maturity of the main energy-intensive industries both decrease the speed of diluting or replacing existing capital stocks with more energy-efficient versions. This means that the rate of improving energy efficiency in the FRG is likely, other things being equal, to be considerably slower than in countries with faster population growth or more dynamic industrial construction. While this does not directly affect our analytic results, which are static rather than dynamic, it is relevant to assessing achievable rates of implementation.

The following analysis uses 1973 as a base year for these reasons:

- The 1973 data reflect the "pristine" state of energy efficiency before the first "oil shock," and thus predate significant efforts to save energy. Between 1973 and 1980, West Germany's GNP grew by 20.4% in real terms while primary fuel demand increased by only 2.7%. It is not known what fraction of this energy saving is due to efficiency improvements and what to behavioral changes of such a character that they were not reflected in reductions of GNP, although the analysis below suggests that even if the entire improvement were due to technical gains, they would still barely have scratched the surface of what is worth doing.[6] Using a base year more recent than 1973 would compare technical potential with a moving target and risk double-counting recent improvements.
- Because of the 1974–75 world recession and slow recovery since then, present FRG economic activity levels are only moderately above those of 1973, as noted above.

---

ponds (*New Shelter* 1980) can do the rest. One of us (ABL) has seen a 20°C completely passive house in 40°C tropical heat.

6. Conversely, if the gain were all behavioral, that would imply an even greater scope for technical improvement not yet begun.

- The data for 1973–74 have been particularly well analyzed in the German literature. There is often a lag of several years in getting recent data.

Tables 2.1 and 2.2, therefore, show basic energy data for the FRG in 1973, disaggregating end use into 15 sectors and uses. The conventional unit, 1 million metric tons of hard coal

Table 2.1 *Energy use in the FRG, 1973*

Category	MTCE	EJ ≡ 10^{18}J
Primary energy, comprising:	378.5	11.09
feedstock	29.9	0.88
fuel	348.6	10.21
Primary fuel, comprising:	348.6	
coal	115.0	
oil	184.2	
natural gas	35.5	
other (nuclear, hydro, wood, etc.)	13.9	
End-use energy, comprising:	253.9 =	0.73 × primary energy
electricity	30.6	(12% of end-use)
liquid fuels for transport	47.8	(19% of end-use)
other fuels and heat	175.5	(69% of end-use)

Population: 61.98 million
Primary fuel use per capita: 5.216 kW

SOURCE: AG Energiebilanzen 1973.

Table 2.2 *Structure of energy end-use in the FRG, 1973*

Sector and application	End-use energy (MTCE)			% of sector	% of total
	fuel	electricity	total		
Space heat, comprising:			90.0	82	35
residential	55.1	1.9			
commercial[a]	32.5	0.5			
Process heat, comprising:			13.4	12	5
residential[b]	5.4	1.9			

(Table 2.2 *continued*)

Sector and application	End-use energy (MTCE)			% of sector	% of total
	fuel	electricity	total		
commercial[a]	3.3	2.8			
Appliances[d]			6.0	6	3
residential		3.6			
commercial[a]		2.4			
TOTAL residential & commercial[ac]	96.3	13.1	109.4	100	43
Automobiles	27.2			55	11
Trucks & buses	10.7			21	4
Rail, ships, & miscellaneous transport[e]	6.4			13	3
Aircraft[f]	3.5			7	1
Electric rail, subways, trams		1.1		2	
TOTAL transport	47.8	1.1	48.9	100	19

		electricity			% of	% of
	fuel	process	other[k]	total	sector	total
Steel	29.3	0.4[g]	2.1	31.8	33.3	13
Chemicals	10.5	3.0[h]	1.8	15.5	16.2	6
Cement	5.3		0.3	5.6	6.0	2
Other industries: electrical		2.0[hi]	6.8	8.8	9.2	3
Other industries: process heat	25.0			25.0	26.2	10
Other industries: space heat[j]	9.1			9.1	9.5	4
TOTAL industrial[m]	79.2	5.4	11.0	95.6	100.	38
TOTAL, ALL SECTORS[m]	223.3	30.6		253.9		100

SOURCES: AG Energiebilanzen 1973; Reents 1977; Suding 1980.
a . Including military.
b . Residential hot water & cooking.
c . Excluding military and agricultural transport.
d . Electrical uses except space and water heating.
e . Includes coal and diesel for rail, gasoline and diesel for military and agriculture.
f . Including military aircraft.
g . 5 million metric tons electric steel-making @ 600 kWe-h/T.
h . Electricity-specific processes and process heat.
i . Includes aluminum production.
j . Covering 85% of industrial employees.
k . Lighting & drives.
m. The 0.2 MTCE discrepancy with the actual sum of the entries is due to rounding errors.

equivalent (MTCE in English, Mio tSKE in German), is equivalent to 29.3 PJ (29.3 × 10^{15} J), 27.8 trillion BTU, 8.14 TW(t)-h, 5.05 million barrels crude oil equivalent, or 7 trillion (7 × 10^{12}) kcal. Both tables are summarized graphically by Figure 2.1.

To explore improvements in the 1973 efficiency levels, we shall seek to specify each activity level (including its per capita value) and the practically achievable improvement in "technical coefficient": that is, the energy intensity of providing a physical unit of that good or service under conditions characteristic of the FRG, according to cited literature. This improvement will be shown as an index relative to 1973 levels (defined as 1.00). Thus an index of 0.6 means that the 1973 FRG level of that energy service can be supplied with 0.6 of the energy shown in Table 2.2 (i.e. with 40% less energy).

To show the degree of conservatism in each assumption about technical potential, we shall broadly characterize the early 1981 status of technologies as commercial, demonstrated and available, or demonstrated and under engineering development to reduce to wide commercial practice. Why do we consider technologies which are entering commercial service as well as those which are in commercial service? Because market imperfections (such as inequitable access to capital and information, or split incentives between those in a position to make energy-saving investments and those who benefit from the savings) and the time lag between increasing fuel prices and technological adjustments together mean that many technologies for saving energy are available but not yet commercially used on a large scale. Many more are technically straightforward but have not yet undergone the needed final engineering development for mass application. Consequently, prices cannot always be obtained from the marketplace, but must also in some cases rely on engineering analysis and estimates reflecting normal industrial and commercial practice. We have not, however, based our conclusions on future technological improvements, even though these are occurring very rapidly, but only on devices already in or entering commercial service. Most, though not all, of these devices are available—and all of them could easily be both made and sold—

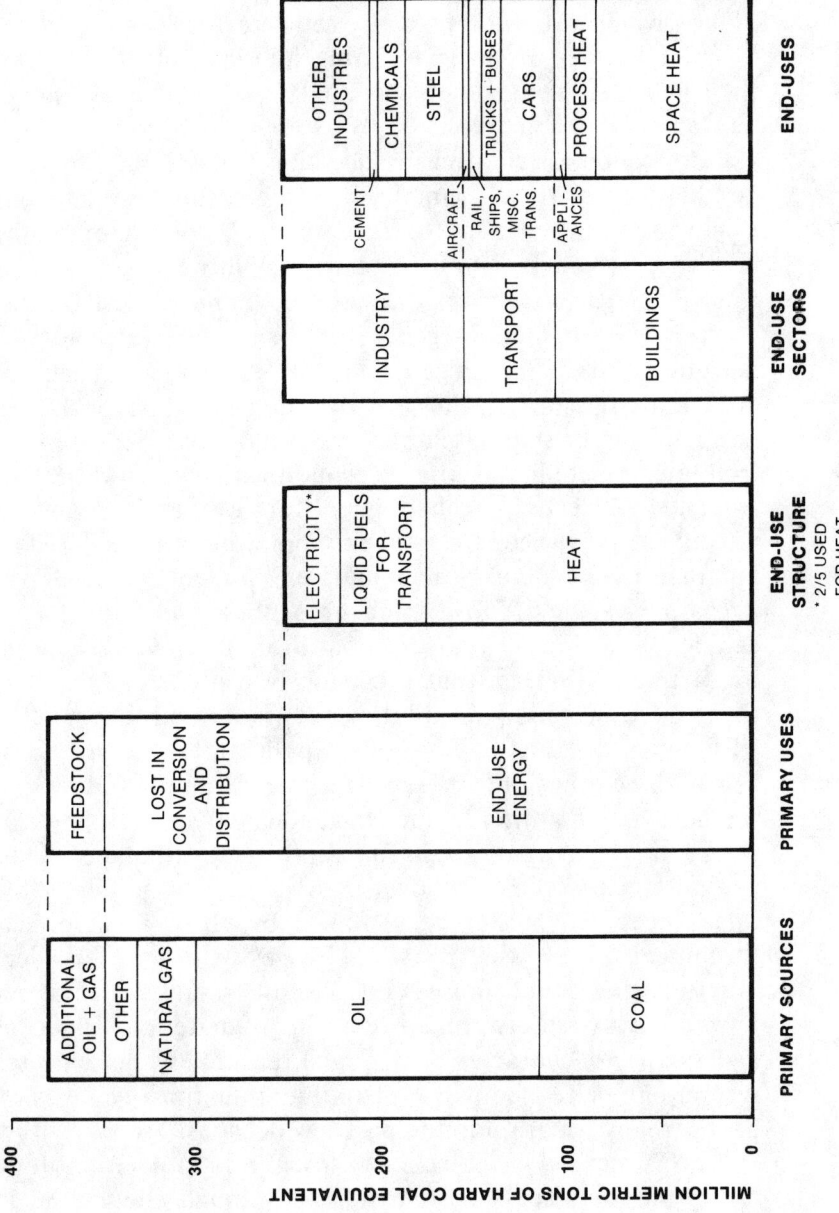

Figure 2.1 *Patterns of energy use in the Federal Republic of Germany, 1973*

in the FRG, though some are still in the process of diffusing into that market. Some of our efficiency gains are based on techniques which are widely used elsewhere (such as the Western Canadian styles of superinsulating houses) but not yet used in the FRG; we see no reason why analogous techniques should not be transferable and adaptable to German conditions.

Our economic criterion for selecting technologies is that they be able to provide the same service more cheaply than conventional long-run marginal sources: namely, nuclear or synthetic-fuel systems ordered now. Where possible we give the payback times for alternative investments based on *present* FRG fuel and power prices (which in general are substantially below those marginal costs [Clausen & Franke 1979]). These payback times are mainly far shorter than the several decades' payback for new power plants or synfuel plants; so our assumed measures are far from a complete list of what is economically worthwhile.

Besides selecting technologies for availability and cost-effectiveness, we need to make one somewhat more subtle type of judgment: how quickly technologies can be deployed on the necessary scale. Those measures that we consider fully achievable within 50 years without dramatic intervention in the marketplace—indeed, that might do best if allowed to compete freely—we have assembled into a snapshot of an energy-efficient FRG marked "2030." For some applications we give an additional value labelled "[present] technical limit." This refers not to the need to wait for additional technical development, but rather to the need to wait for complete overhaul of a major capital stock which turns over slowly, such as buildings, where this process might not be complete by 2030. In a few cases, "technical limit" also refers to changes in technical devices or practices that could imply some modest behavioral adjustments by end-users (notably greater recycling of materials) which are of a degree and character that could reasonably be expected in response to price, and which are milder than those changes which have already occurred in the past few decades.

By this method we calculate below the total potential for efficiency improvements by showing how much energy the FRG economy would have required in 1973 had these improvements been in place then. Thus we ignore the structural changes in the mix of energy services that are likely to occur in the future. Growth scenarios based on present saturation levels and on structural trends observable in affluent countries can be constructed to examine these effects, and where this has been done, as for the FRG (Krause et al. 1980), it has shown a shift of end-use needs which only strengthens our conclusions.

Finally, our emphasis here on engineering-economic analysis should not obscure the substantial energy savings already being achieved by changes in lifestyle, which in the FRG may represent an end to carelessness in some cases, a natural evolution of values towards a "conserver society" (Science Council of Canada 1977) in others, and a social hardship in still others. For example, during the year 1980 alone, West Germans in oil-heated homes decreased their average residential energy use by 13%. Most of this large and quick reduction resulted from lowering thermostats and conditioning fewer rooms. Our assumption in this paper—that people throughout the world will comfortably condition 100% of their new floorspace (assumed to rise to present FRG levels per capita) when this level of comfort was not even achieved in the world's wealthiest countries in pre-oil-crisis days—is only one of the many conservatisms which we tacitly include by neglecting here the potentially important role of values and lifestyles (CONAES 1980; Carlson et al. 1980).

With these caveats in mind, we now examine in turn the 15 end-use categories shown in Table 2.1.

2.3.1. RESIDENTIAL AND COMMERCIAL SECTORS

Energy use in the FRG's built environment is dominated by space heating (82%), followed by process heat and hot water (12%) and electricity-specific uses, notably lights and small appliances (6%).

Residential space heat. In 1973, the FRG's occupied floorspace per capita averaged 25 m^2. The average consumption of end-use energy per Celsius degree-day (DD) and per unit of occupied floorspace, based on an average of 3300 degree-days per year (referred to 18°C), was 312 kJ/DD-m^2. Since the average end-use conversion efficiency of heating systems was 52% (Reents 1977), the average use of heat at the room register was 162 kJ/DD-m². Based on typical pre-1973 construction practices (Dittert et al. 1978), the average heat-loss coefficient in the FRG housing stock was 300 kJ/DD-m². This average includes a range of types, vintages, and construction practices, and that range has been disaggregated in detail by studies in the UK (Romig & Leach 1977), Denmark (Nørgård 1979), and the FRG (Dittert et al. 1978; Krause 1980).

New buildings. The state of the art for new residential buildings is represented by "superinsulated" or "micro-energy" houses, whose annual space-heating energy requirements are less, often by severalfold, than their water-heating energy requirements (Rosenfeld et al. 1980; Shurcliff 1980; U. of Saskatchewan 1979). They feature wall and roof U-values of 0.08–0.15 W m^{-2} C°ΔT^{-1}, double- or triple-glazed windows with insulating night shutters, tight vapor barriers, infiltration rates of <0.05–0.1 air changes per hour, and ample ventilation via air-to-air heat exchangers which ensure good indoor air quality but recover about 80% of the heat (or coolness) in the outgoing air (Besant et al. 1978; Rosenfeld et al. 1980). (See also Table 4.1 below.)

This construction practice lowers gross shell heat losses so far that internal heat gains from occupants, cooking and appliance use as well as solar gains through windows suffice to establish normal comfort levels without supplementary heat whenever outdoor temperatures exceed −5°C to +1°C in cloudy weather (Rosenfeld et al. 1980; Dumont 1981). The best heat loss value so far, measured with a thousand heat-losing visitors per week, is 39 kJ/DD-m² in the Saskatchewan Conservation House, at latitude 50.5°N in Regina. Internal and solar gains reduce this to a net heat requirement at the room of only 4.5 kJ/DD-m² (Besant et al. 1978), or 5.1 GJ/y for a 187-m² detached single-family house in a 6000 DD/y climate—nearly twice as cold as the FRG—with an average January outdoor temperature of −18°C.[7] This

---

7. Rosenfeld et al. (1980) give 13.2 GJ/y net (11.7 kJ/DD-m²) based on 1978–79 data

remarkable performance, achieved without a special design to maximize passive solar gain (1–2 m² of extra south window area could have eliminated the net heat requirement), lets one conclude with confidence that in a country like the FRG, which has a milder climate and a large share of multi-family dwellings (with fewer outside walls and larger volume-to-surface ratio), the use of these building practices would result in houses needing no active space heating system whatever, or at worst a slight oversizing of the hot water system. In confirmation, performance broadly on these lines has recently been achieved in Alaska and central Norway.

More than 250 superinsulated houses have been built with Saskatchewan-style techniques in the US and Canada. They have a range of net space heating requirements of 13.2–33.6 kJ/DD-m² (Rosenfeld et al. 1980; Dumont et al. 1980).[8] At least six tract-housing contractors in western Canada are building them routinely in large numbers. The Canadian Department of Public Works is making a major effort to spread the technology. By the end of 1982, another 1000 such houses are scheduled to be built in 100 Canadian communities. The technology is also starting to enter the West European prefabricated housing market. It can be considered demonstrated, available, and near-commercial (US Department of Energy 1981).

From Canadian contractors' experience, the additional costs of such superinsulated construction, compared with an otherwise equivalent standard house, are (1980 C$ = US$0.83): infiltration reduction, C$500; air-to-air heat exchangers and mechanical ventilation system, C$800; superinsulation in walls, floor, and roof, C$2800; triple glazing, C$500; total, C$4600 (US$3820). If no furnace is required, however—as is generally the case even in the Canadian climate with good construction—there is a C$1500 saving from not having to install a gas furnace, so the net cost is C$3100 (US$2570). Even the C$4600 gross

---

when shutters leaked and more than 1000 visitors per week lost extra heat; the designers expected to get this back down to about 5 GJ/y with "tuning up."

8. The higher figure is inferior to measured performance by Illinois Lo-Cal houses with negative marginal capital cost (Shick 1979; Leger & Dutt 1979).

marginal investment, about 7–8% of standard construction costs, pays back against Canadian natural gas (or electricity) prices of $2.5 ($7.5)/GJ in 16 (5) years (Dumont 1981). At current FRG prices of about $7 ($18)/GJ, assuming a similarly large house in the milder FRG climate, the payback time would be 10 (4) years, or about 7 (3) years with the furnace saving. Alternatively, a comparison of capital cost alone, under FRG conditions, for new construction of superinsulated instead of standard houses *vs.* marginal nuclear electric-resistance (or electric heat pump) space heating shows that the former option has a capital-cost advantage of about a factor of four (two) (Krause et al. 1980a). Real construction costs for superinsulated houses, using a housing construction deflator, are actually falling from the indicated levels as contractors gain experience—some total marginal costs are less than C$2500—and the air ventilation system, currently made in small lots, is expected to fall in installed price from C$800 to C$400 with wider commercialization.

Another important feature of superinsulated houses is that they make solar water heating and hybrid solar space heating far more attractive economically, as noted in Section 4.2.1. We do not include this feature in calculating the economics of energy saving, though it can reduce the cost of delivered heat from the solar system by about an order of magnitude, and this type of synergism is important to our later consideration of solar heat economics.

We conservatively take these single-family house performance data rather than an average mix as the basis for the technical potential in all new dwellings (including multi-family dwellings), and assume an average new house value of 8 kJ/DD-m^2 of net space heat required at the room—almost twice what the best Saskatchewan house actually needs.[9] Compared to the status quo in the 1973 building stock, 162 kJ/DD-m^2 (partial heating of floor space only), this yields an index of 0.05.

Existing buildings. Existing buildings can be retrofitted

---

9. Or about the same as Rosenfeld et al.'s (1980) value without their assumed 0.7 furnace efficiency; passive heat can more than make up the difference.

(modified) with a wide range of techniques, such as outside and/ or inside wall insulation including vapor barriers, weatherstripping, triple-glazed windows or (usually cheaper and more effective) insulating window shutters and shades (Shurcliff 1980a),[10] roof and basement insulation, mechanical ventilation with heat exchangers, added greenhouses or winter gardens, and glazing of exterior masonry walls to form a passive-solar Trombe wall. All these techniques work well in cold, cloudy climates: in New England, for example, over half of all the solar systems now in use are passive, and over half of those were made by adding passive solar features (mainly south-facing greenhouses) to existing buildings. In areas with frame buildings, notably western Canada, an experienced contractor can reportedly remove wood siding, install a vapor barrier and superinsulation, extend window and door casings and eaves, and restore the siding for US$3000–4000 above residing-only costs (Palmiter & Miller 1979); this technique can readily save upwards of 90% of the space heating load, paying back in a few years (Bryenton 1980). Even if the house were initially insulated to Canadian electric heating standards—about 182 kJ/DD-m^2 (Schipper & Ketoff 1980)—the retrofit would cost only C$104/GJ-y saved, or about half the price of FRG electricity today (ibid.).

In principle, then, the same technologies can be used for existing buildings as for new ones. In practice, in the entire building stock somewhat lesser average efficiencies are to be expected because of the difficulties of removing bypass and natural infiltration and heat bridges in old buildings, because of the need to protect historic buildings, and because of variable quality of workmanship. Retrofit insulation of wall, roof, and basement insulation, multiple glazing, window insulation, and weatherstripping are commercial technologies in the FRG. Mechanical ventilation systems with heat exchangers are also available but need to be better adapted to residential buildings.

The empirical package costs of thorough retrofitting range

---

10. Another attractive option is the range of new glazing materials such as Heat Mirror™, now in mass production and offering a U-value of about 0.6, and glass-and-plastic sandwiches containing insulating gases such as argon, some offering U-values of about 0.3.

from $100 to $200 per square meter of floorspace in the FRG, assuming DM 1 = $0.5 during 1978–80 (Dittert et al. 1978; Hake 1980; Krause et al. 1980). Although these costs are some 5–10 times higher than equivalent US costs (Rosenfeld et al. 1980; ACEEC 1981) and several times higher than in the UK (Romig & Leach 1977), they are nevertheless attractive against marginal nuclear direct or heat pump systems (Krause et al. 1980; Hake 1980). In many cases, FRG retrofits can even compete against present average FRG fuel prices—for example, heating oil at $9/GJ, or $13/GJ at the room. A retrofit contest sponsored by the FRG government in 10 apartments saved an average 46% of heating fuel, cost-effectively at 1977 fuel prices and with construction times of 3–6 weeks (Therma-Wettbewerb 1977). Similar benefits were shown in the Ulvsunda project in the Stockholm suburb of Bromma, where apartment retrofits in several stages cut heating oil use by 47% (soon to be 58%); all the measures used, even especially costly techniques (e.g., exterior insulation with a new rendered façade for esthetic reasons), were cost-effective at 1980 oil prices (Höglund et al. 1981).

An average reduction of heat losses in the existing stock by 80% (Krause 1980) yields energy intensities (net heat requirements at the room after accounting for "free heat" from passive solar gain, lights, appliances, and people) of about 30 kJ/DD-m^2 if all occupied floorspace is heated, which is not nearly the case today. The relative energy intensity compared to the present 162 kJ/DD-m^2 is thus 0.185. By the year 2030, a building mix of 80% old and 20% new (post-1981) buildings can be expected (ibid.). The average energy intensity of the stock in 2030 would then be 26 kJ/DD-m^2—in relative terms, 0.16.

Heating systems. In the early 1970s, 88% of all residential heating systems in the FRG were direct-fuel systems. Central heating had a 62% share; single-room systems, 38%. The average end-use conversion efficiency was 52% for all heating systems and 48% for direct-fuel systems (Reents 1977).

Heating efficiencies can be improved in new or old buildings by exhaust-stack lids, outdoor and indoor electronic sensors and thermostats, thermostatic valves, electronic fuel-air mixing regulators, smaller peak furnace capacities for better load matching (so that less fuel is wasted heating up the mass of the furnace itself each time it goes on), insulation of distribution pipes and ducts, and regular maintenance. Pulse-combustion furnaces with 90–95% First Law efficiency are available. Other options include new or retrofit district heating systems, total energy systems with local electrical generation and waste heat distribution, and solar heating systems, either single-building or district (Margen 1980; Gleason 1981). All these technologies, with the possible exception of solar district heating, are commercial.

The simpler retrofits (those other than solar, cogeneration, and district heating) are cost-effective against historic oil and gas prices of \$5/GJ, with payback times of 1–5 years (Dittert et al. 1978). An exception is insulation of distribution ducts, which is cost-effective against present FRG oil and gas prices (\$9/GJ). New solar heating systems in the FRG are cost-effective at present fuel prices if operated "bivalently" (with fueled backup) (Schürle et al. 1977) and are cost-effective at the margin in a "monovalent" (stand-alone) system in superinsulated houses (Krause et al. 1980). District heating systems based on diesel engines or steam plants are used commercially in the FRG and deliver heat cost-effectively against present fuel prices (Hein 1979).

With these heating technologies, an average end-use conversion efficiency of at least 80% can be achieved in the FRG. The relative efficiency of 1973 as against improved heating systems is thus $0.52/0.80 = 0.65$. To supply the efficient building stock with its 26 kJ/DD-m^2 at the room, the improved heating systems will require no more than 32.5 kJ/DD-m^2. Relative to the 1973 end-use energy intensity of 312 kJ/DD-m^2, this yields an index of 0.11.

Commercial space heat. The commercial sector contains buildings as diverse in function as offices, schools, hospitals, hotels, and small factories. Its heat loss characteristics are similar to those of multi-family dwellings in the residential sector (Dittert et al. 1978).

Remarkable progress has been made in commercial energy efficiency. Several new high-rise, heavily glazed office buildings in Canada do not require any active space heating at all: on the contrary, they need cooling even in mid-winter. (Further design refinement can probably balance out that load, too.) Internal gains from lighting, people, and equipment suffice if the heat is properly distributed with an air circulation and heat storage system. The cooling is done similarly with a self-ventilating façade (Encon 1978). These designs pay back their marginal construction cost in a few years at present Canadian energy prices, which are very much lower than those in the FRG.

In existing FRG commercial buildings, retrofits to recover heat and improve heating systems show the best payback times (<1–5 y) (Dittert et al. 1978). Most retrofit techniques for heating systems and building shells are analogous to those in the residential sector. Retrofits done by private energy consulting firms in the US have yielded 30–60% energy savings in commercial buildings with payback times of a few years against present US fuel and power prices; in well-designed programs, the price is about $10/m^2 of floorspace or $1/GJ (Rosenfeld et al. 1980; ACEEE 1981; SERI 1981). Similar findings for the FRG are reported in the very detailed Battelle study (Dittert et al. 1978). Much larger efficiency improvements are attractive at marginal fuel prices.

Assuming a 75:25 ratio of old to new commercial buildings in the FRG by 2030 and a relative space heating energy intensity of 0.10 for new and of 0.25 for old buildings, an average relative end-use energy intensity of 0.21 can be achieved with improved designs and heating systems. The long-term technical limit is taken as 0.05 (Krause 1980; Rosenfeld et al. 1980).

Residential process heat. The average daily FRG consumption of hot water was about 37 liters per person in 1973, corresponding to an end-use energy demand of about 3 GJ/cap-y at an average conversion efficiency of 66% (Reents 1977). The efficiency of this service can be improved by heat recovery from waste water (Orth 1976), insulation of storage tanks and pipes, shower heat switches, and improved heaters.

Heat recovery from residential waste water is a demonstrated and available technology. Recovery of greater than 50% is feasible (Orth 1976); FfE & RWE (1977) show 89% direct recovery without a heat pump. The other measures just mentioned are already commercial, and new heating systems with higher summertime efficiencies than oil-fired furnaces are also commercially available (e.g., district, diesel cogeneration, and solar hot water systems). An 83%-First-Law-efficient gas water heater, including standby losses from the tank (which many FRG water heaters do not have), has been demonstrated in the U.S. This can be further improved by preheating domestic water with refrigerator or freezer waste heat.

Payback times for shower heat recovery systems would be less than three years (Orth 1976). Extensive heat recovery from graywater would save 1800 kW-h of end-use energy per household in the FRG at a cost of about $10/GJ (FfE & RWE 1977)—cost-effective against present FRG electricity prices and, due to the low summer efficiency of oil water heating, also against present oil prices. All other technical measures listed above are cost-effective against present energy prices, as noted earlier in discussing space heat.

With an average heat recovery of 50% in the building stock and an end-use conversion efficiency of 85%, the index of energy intensity would be 0.40.

Commercial process heat includes water heating, cooking and baking, drying, heating of construction materials, and space-heating greenhouses. Greenhouse heat requirements can be reduced by at least 80% through triple or heat-reflective glazing, moveable insulation, and more efficient heaters. (Experience in some formidable North American climates suggests that with adequate thermal mass and modern glazing materials, entirely passive heating is not difficult to obtain.) Water heat recovery can be very high in many commercial installations, where efficient water heaters, automatic shut-off valves, and insulation of distribution ducts also apply. End-use energy requirements for cooking can be lowered by 54%, and for drying by up to 59% (Nørgård 1979). Because an estimated half of the total FRG demand classified as commercial process heat is for heating greenhouses—a task now done in considerably colder and cloudier climates with no energy at all—a reasonable relative energy intensity averaged over this category, using best cost-effective technologies, would be 0.3.

Residential electricity-specific functions. Table 2.3 shows the 1973 degree of saturation in FRG appliance ownership per household. A 100% saturation in all categories would result in a

Table 2.3 *Cost-effective electricity savings in household appliances*

Appliance	1973 FRG Saturation[a] (100%=21.8M)	Est. Av. 1973 FRG Use,[a] kW-h/unit-y	Improved kW-h/unit-y[b]	y payback @ 5¢/kW-h[b]
refrigerator (220ℓ)	87%	400	90[c]	9
freezer	32%	750	145	8
washing machine	80%	450	71	6
dishwasher	7%	880	95	8
clothes dryer	4%	440	130	9
furnace pumps & fans	45%	300	140	4
TV, miscellaneous small items	100%	300	(unchanged)	0
lighting	100%	250	117	0
electric cooking	64%	600	440	7
total kW-h/y per 100%-saturated household		4370 = 2.86 × 1528		average 6.3 (use-weighted)

a. Data from Krause (1980).
b. Data from Nørgård (1979, 1979a); lighting @ 2.84 persons/household and with per capita lighting use corrected to 1973 FRG levels. The payback times shown are for the last (costliest) increment of efficiency improvement to the level shown, and considerably exceed the *average* of all measures used to achieve that level.
c. This specific consumption of 0.034 kW-h/ℓ-month compares with 0.13 expected for a vertical model being built by Larry Schlussler (725 Bayside Rd. #6, Arcata CA 95521), 0.053 measured for his horizontal prototype (*Soft Energy Notes* **4** (3):95-96 [June/July 1981]), 0.11 for good 1981 Japanese models, 0.12 for the best 1981 US model (Amana ESTR-18D), and ≳0.23 for the average US frost-free unit in use in 1980. A seasonal-storage icebox would provide superior refrigeration with no electricity, but is not included here. A further conservatism is our use of linear scale-up in normalizing various unit sizes.

specific commercial-sector uses we conservatively assume a relative energy intensity of 0.33.

Summary of residential and commercial sectors. Table 2.4, summarized by Figure 2.2, shows that with available and cost-effective technologies, 82% of the FRG's energy end use in the built environment can be eliminated within a few decades without affecting the quantity or quality of services provided. The difference in the indicated technical potential between residential and commercial space heat entries stems from a more conservative treatment of the commercial sector. This arises from a less thorough data base for commercial than for residential buildings, and probably does not represent a real technical difference: on the contrary, such data as are available on retrofits in the US and elsewhere (ACEEE 1981; SERI 1981) suggest that the savings in commercial buildings are if anything usually larger and cheaper than in residential buildings.

2.3.2. TRANSPORT SECTOR

Cars. About 17 million cars were registered in the FRG in 1973, corresponding to 380 cars per 1000 adults. Average fuel consumption was 10.6 ℓ/100 km or 22.5 mi/US gallon; 97% of engines were gasoline Otto; and average power-train efficiency (from fuel to wheels) was 10–15%.

The technical options for improving vehicle fuel efficiency have been extensively reviewed in several recent publications (Gray & von Hippel 1981; SERI 1981; *Soft Energy Notes* 1980; Shackson & Leach 1980; Seiffert et al. 1979, 1980; TRW 1979; FfE 1977). These options can be grouped into four main areas:

- power train optimization;
- reduction of weight and rolling resistance;
- reduction of aerodynamic drag; and
- downsizing (shifting to smaller cars).

The percentage savings from each measure of each kind are not strictly multiplicative; the rules for combining several savings without double-counting are discussed elsewhere (Gray & von Hippel 1981; SERI 1981; Ross & Williams 1981). (We count

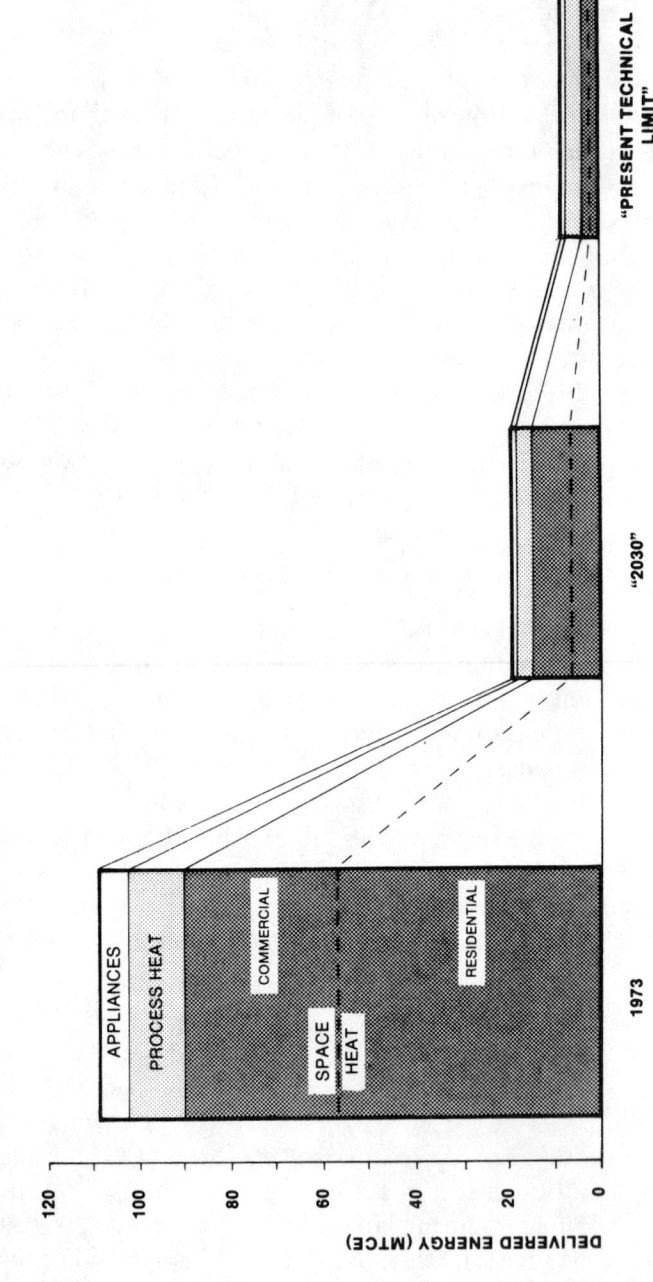

Figure 2.2 Cost-effective potential efficiency improvements in the FRG residential and commercial sectors, at constant 1973 activity levels

here only more efficient cars, not modal shifts to public transport.)

Available improvements in the power train (TRW 1979) are: improved lubricants for engine and drive-train components (5% fuel savings); Xylan™ coatings in engines (10–15%); replacement of conventional gasoline Otto engines with stratified-charge low-RPM engines (25%) or with direct-injection turbocharged diesel engines (30%); continuously or infinitely variable transmissions, which enable engines to run at an optimal constant speed (20%); such transmissions combined with regenerative braking and energy storage (35%); idle-off systems (12%). The power-train efficiency can be doubled by a variety of combinations of these technical measures and can even be tripled by combining a turbocharged diesel engine with a continuously variable transmission, regenerative braking and energy storage, idle-off system, Xylan™ coatings, and improved lubricants (TRW 1979). (Such measures as microcomputer engine control are taken for granted.)

Weight can be reduced by use of composite materials and energy-absorbing materials (foam-filled metal structures or metal foams). Further weight can be saved by optimizing the design dimensions for a given interior volume. The 1973 German car offers plentiful opportunities for these measures, particularly the former: of its total weight (approximately 1000 kg), plastics contributed less than 5%, nonferrous metals another 4–5%. Weight savings of 25–40% are feasible without reducing vehicle size.

Tire rolling resistance can be decreased by 20–30% below that of bias-ply standard tires by using conventional radials, resulting in a fuel saving of about 5% in the driving cycle. New radial tires of improved design are now being marketed which offer a further 3–4% increment, and more advanced radial technology allows another 5% (SAE 1980). Conventional radial tires have been standard equipment on new FRG cars for many years. Advanced radials, used consistently, would save 8–9% of car fuel.

Aerodynamic drag coefficients in the 1973 FRG car fleet ranged from 0.35 to 0.69 (FfE 1977), averaging about 0.5—about three times the theoretical minimum. Values of 0.24–0.30 can be reasonably achieved (Seiffert & Walzer 1980), saving 10–12% of fuel.

The 1973 FRG car fleet consisted overwhelmingly of four-passenger models, even though the estimated average passenger load was only 1.9. Most trips require no more than two passenger seats. An attractive way to achieve fleet downsizing, then, is somewhat greater vehicle specialization: slightly decreasing four-passenger vehicle size, but also shifting to an increased fraction of two-passenger vehicles in the fleet. Shifting from an optimized four- to two-passenger design saves up to a third of the fuel (SERI 1981).

In combination, the above technologies can lower vehicle fuel consumption by about 80% to 2.1 ℓ/100 km (110 mi/US gal). With a 30% share of two-passenger cars in the fleet, this would drop to 1.9 ℓ/100 km (125 mi/US gal). Yet no technical breakthroughs are required for any of these vehicle technologies. Turbocharged diesel engines, stratified-charge engines, improved lubricants, and improved radial tires are now commercially available. Continuously variable transmissions will be marketed starting in 1982 (DAF/Borg-Warner/Fiat) for smaller cars, and simplified designs are being rapidly developed. Composite materials are now being used for some components of commercial production vehicles. Xylan™ coatings, improved stratified-charge engines, and energy-storage transmissions are now under advanced development and have been tested in prototypes. By combining many of these elements, Volkswagenwerk AG has already achieved an on-road average driving cycle fuel consumption of only 3 ℓ/100 km (80 mi/US gal) in a four-passenger car comparable to the Golf (Rabbit) (Seiffert & Walzer 1980); recent US tests of a VW prototype yielded EPA efficiencies of 80 mi/US gal city, 100 highway (3.0–2.4 ℓ/100 km). Several manufacturers are developing prototypes with vehicle weights as low as 510 kg (MiniMetro) to 650 kg (VW Golf). Some of these cars use advanced impact-absorbing materials to withstand crashes better than standard cars do.

The costs of automotive redesign have been reviewed in detail (Gorman & Heitner 1980; Shackson & Leach 1980; SERI 1981;

von Hippel 1981[11]). The fuel costs at early-1981 prices in the FRG ($0.65/$\ell$ @ DM 1 = $0.5) are roughly equivalent to $18/GJ. The major improvements above cost $3–5/GJ for turbocharged diesel engines, <$4/GJ for Xylan™ coatings and improved accessory drives (Gorman & Heitner 1980). For an improvement from 10.6 ℓ/100 km (22.5 mi/gal) to 4 ℓ/100 km (60 mi/gal), estimates of total cost (in 1980–81 dollars) range from $800 (Gorman & Heitner 1980) to $1500 (von Hippel 1981). These extra costs are well below those consumers should be willing to pay at a crude oil price of $40/bbl ($6.9/GJ), let alone at the 160% higher FRG gasoline price. In fact, at the average 1973 FRG driving distance of about 12,700 km/car-y, an improvement to 4 ℓ/100 km saves each driver about 840 ℓ/y, costing at present FRG prices about $545/y or (assuming that real gasoline prices escalate only at the real discount rate) a present value of $5450 over a nominal 10-year car life. Clearly, even von Hippel's conservatively high estimate represents a payback time of less than three years to an average FRG driver today. It follows that improvements even beyond 4 ℓ/100 km (60 mi/gal) are also economically attractive, and insofar as this is done by downsizing or increasing the fleet fraction of two-passenger cars, this would actually reduce the cost of the car.

For the "2030" case, we assume a fleet of four-passenger cars only, using best available technology. This yields, conservatively, 2.4 ℓ/100 km (100 mi/US gal) or a relative energy intensity of 0.23 relative to 1973. Increasing the share of two-passenger cars yields a fleet average of 1.9 ℓ/100 km (125 mi/US gal), a relative energy intensity of 0.18, as our "present technical limit" case.

Trucks and buses. Of the 10.7 MTCE indicated for these vehicles in Table 2.2, 9.9 MTCE in 1973 was used by freight trucks. Total freight transport by road was 94.9 billion ton-km in 1973, of which 41% was short-distance and intra-urban hauls of <50

---

11. As quoted by SERI (1981:Table 14a); we subtracted from von Hippel's $2263/car total cost the increment associated with improving from the present US fleet average of 16.7 mi/gal to the 1973 FRG average of 22.5 mi/gal.

km. Average fuel use over all hauls was 3.1 MJ/ton-km. The majority of FRG trucks and buses have diesel engines and radial tires.

With due allowance for different weights, the same basic menu of efficiency gains applies to trucks and buses as to cars. Diesels can be turbocharged (5–10% saving), Freon® turbo-compounded (15%), or designed for adiabatic operation (10%). Transmissions can be controlled by microcomputers (14–20%). Regenerative braking and energy storage could reduce fuel consumption by 20–35%. Advanced accessory drives (including changes as simple as fan clutches), Xylan™ coatings, and improved lubricants also apply. Conversion to radial tires with advanced low-loss design and replacement of double tires with wide singles (Giles 1979) would save 9–23%. Aerodynamic drag can be reduced by up to 50% with "slippery" front-end designs, resulting in a 20–25% fuel saving in long-distance hauling (Steers & Saltzman 1977).

All these technologies are demonstrated. No breakthroughs are required. An urban bus with regenerative braking and energy storage has been prototype-tested in the FRG. Further engineering development is desirable in some applications, but should present no difficulties. Radial tires, turbocharging, and streamlining retrofits are in wide commercial use today and are cost-effective at present fuel prices (cab air deflectors often pay back in months). As in the case of cars, substantial room exists for further cost-effective improvement. Combining the most cost-effective technical measures with efficiency improvement by somewhat higher payloads through better shipping management (Hann 1979) results in a relative energy intensity of 0.40. We omit possible savings from rail piggybacks and substitution of containerized airships, now being developed in several countries for door-to-door freight.

Fuel for rail and ships. Better load management, lighter rolling stock, regenerative braking, and replacement of steam by diesel locomotives (completed in 1977), and in the case of ships, improved diesel engines and drag reduction with anti-fouling paints, can lower energy intensity 25%. (A Japanese ship has added heat recovery, variable-pitch propeller, and sails, saving nearly 50% [Tsuchiya 1981].) We assign the same potential to military and farm vehicles.

Electric rail, including urban mass transit rail systems. Possible improvements include better electric motors, lighter stock, regenerative braking, aerodynamic drag reduction, better insulation, and better load factors. Their combined effect can lower energy intensity by at least 30%.

Aircraft. The total aircraft fuel consumption (Table 2.2) of 3.5 MTCE is composed of military (18%), German civil aircraft (40%), and foreign carriers (42%). Intra-German flight distances are <1000 km, causing high relative fuel use on the ground and in takeoff.

Technical measures for improving fuel efficiency reviewed by SERI (1981), Wilkinson (1977), and FfE (1977) include use of wide-bodied jets; geared high-bypass turbofan engines; turboprop engines for short hauls; active control technologies to maintain maneuverability and gust tolerance; and weight savings through the use of titanium, high-strength aluminum alloys with good stress-corrosion resistance, and carbon-fiber composites. There is some scope for increasing average payload (through better management and tariffs) and maneuvering planes on the ground with power wheels rather than engine power. There are also more advanced possibilities, notably laminar-flow designs.

The US commercial fleet improved its fuel efficiency from 17.5 passenger miles per gallon (PM/gal) in 1973 to 25 PM/gal in 1979. With complete introduction of the new generation of aircraft (Boeing 757 and 767, DC9-80, advanced L-1011) this is expected to reach 45 PM/gal. New propfan designs planned for 1986 test flights promise a further 15–20% saving without sacrificing speed or quietness (Shifrin 1981). In the FRG, prop-engined aircraft are being reintroduced for domestic Lufthansa flights. These changes are all driven by the dramatic rise in fuel costs—40% of airline operating costs by the mid-1970s and proving unaffordable for many today.

The potential saving from replacing B-727 and B-737 aircraft in domestic flights with new-technology turbofan-prop aircraft is at least 50%, with a similar reduction on long hauls. We therefore take the relative intensity with best available technology to be 0.5, and with longer-term aerodynamic and materials advances (Wilkinson 1977) to be 0.3 ("present technical limit").

Summary of transport sector. Table 2.5 and Figure 2.3 show that with cost-effective and available technologies, transport energy to supply the 1973 mix of FRG services would be 36% of the actual 1973 energy use, and with a changed car fleet mix and more advanced aircraft design, only 32% of 1973 use.

Table 2.5 *Cost-effective potential efficiency improvements in the FRG transport sector*

Application	Relative energy intensity (1973 = 1.00)		End-use energy (MTCE)			
			"2030"		"Technical Limit"	
	"2030"	"technical limit"	fuel	electricity	fuel	electricity
cars	0.23	0.18	6.3		4.9	
trucks & buses	0.40		4.3		4.3[a]	
rail, ships, & miscellaneous transport	0.75		4.8		4.8[a]	
aircraft	0.50	0.30	1.8		1.1	
electric rail	0.70			0.8		0.8[a]
total			17.2	0.8	15.1	0.8
total of fuel plus electricity			18.0		15.9	
index (1973 = 1.00)			0.36	0.73	0.32	0.73
index of fuel plus electricity (1973 = 1.00)			0.36		0.32	

a. As a conservatism we assume no additional savings in the "present technical limit" case.

2.3.3. INDUSTRIAL SECTOR

The disaggregation shown in Table 2.2 was based on three primary materials industries' dominant share of total FRG industrial energy end use: steel, chemicals, and cement together accounted for 56% of the 1973 total.

Steel. Total 1973 production of crude steel was 49.5 million

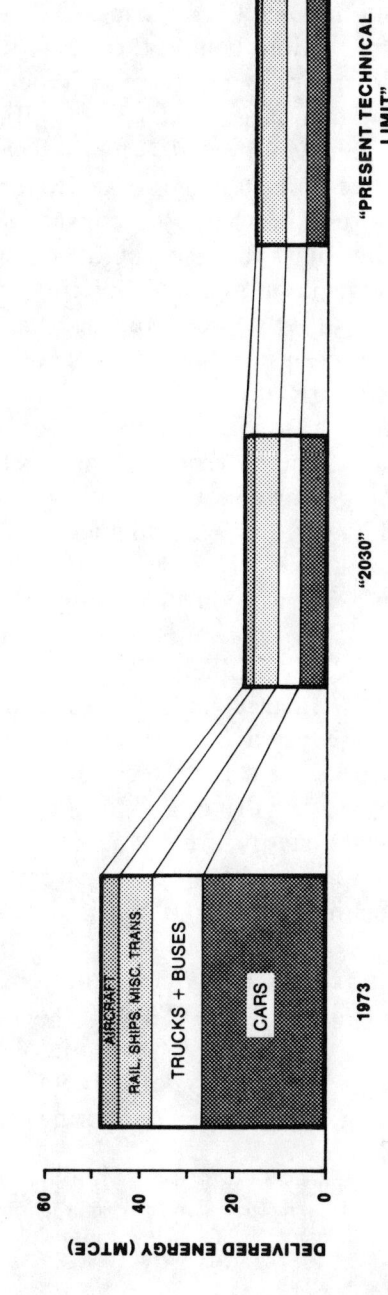

Figure 2.3 Cost-effective potential efficiency improvements in the FRG transport sector, at constant 1973 activity levels

metric tons (MT), corresponding to an apparent consumption of 800 kg/cap. Among OECD countries, the FRG ranked third in absolute steel production, behind the US (136.8 MT) and Japan (119.3 MT), and second behind Japan in per capita production. An international comparison of energy efficiency in the steel industry (Fishelson & Long 1977) has been prepared by the NATO Committee on Challenges of a Modern Society. Such an assessment requires knowledge of consumption data for feedstock preparation, blast furnace operation, steel furnace operation, and slabbing, blooming, and billeting operations, together with the share of different steel furnace technologies, of scrap input, and of continuous casting. Reents (1977) provides the following data for the FRG in 1973.

Ore preparation (13% of steel's end-use energy) was mainly by sintering; 96% of its energy requirement was for fuel, the rest for drives and lighting. For production of pig iron (60% of end-use energy), this split was 94.4% fossil fuels (for both process heat and reductant) and 5.6% drives and lighting. Virtually all pig iron was produced in 1973 with conventional blast furnaces; direct ore reduction was <1.5% of the total. Steelmaking was 67% in basic oxygen furnaces, 22% in older (Thomas and Siemens-Martin) furnaces, and 11% in electric arc furnaces, with 39.5% scrap input. This step required 8% of steel's end-use energy, comprising 77.6% process heat from fossil fuels, 17.6% process electricity, and 4.8% electric drives and lighting. The remaining 19% of steel's end-use energy went to slabbing, blooming, and billeting, and comprised 86% process heat and 14% electric drives and lighting. The output share of continuous casting in 1973 was 16%.

The average primary energy consumption for 1973 steel production in the FRG was 24 GJ/T (FfE 1974), corresponding to a 26% Second Law efficiency[12] (Gyftopoulos et al. 1974). This figure excludes operations in related metal-trade industries and foundries—casting, annealing, tempering, forging, and cold-

---

12. Second Law [of Thermodynamics] efficiency is the minimum energy theoretically needed to accomplish a given change of state, divided by the amount required in practice. First Law efficiency is the conventional useful-output-to-input ratio of an energy conversion device.

rolling—which add another 1.7 MTCE of end-use energy or 1.2 GJ/T of primary energy.

Energy efficiency in steelmaking can be improved by optimizing conventional blast furnace technology and associated processes (ibid.; Fichtner 1977) and by using new direct-reduction processes (Eketorp 1980). Present practice can be improved by optimizing the ratio of iron-ore pellets to sinter in the charge entering the blast furnace; charging hot sinter, pellets, coke and coke-oven gas directly into the blast furnace; preheating combustion air supplied to sinter and pellet furnaces; increasing the air blast temperature and the top gas pressure in the blast furnace; phasing out older steel-making furnaces; injecting oxygen into electric-arc furnaces and preheating the charge; making consistent use of continuous casting; eliminating reheating for secondary rolling; optimizing the use of byproduct gases; and increasing cogeneration of electricity and steam. These improvements can reduce the specific energy requirements for producing steel shapes to 18 GJ/T with a basic oxygen furnace (Gyftopoulos et al. 1974). With a mixture of 75% basic oxygen and 25% electric furnace steel, the average primary fuel requirement can be lowered to 14.8 GJ/T, 40% below the 1973 FRG average. Similar efficiencies are reported for direct-reduction processes: Eketorp (1980) states that 18 GJ/T can be achieved in integrated ore-based steelworks using the Plasmamelt, ELDRED, or INRED processes.

Optimizing energy efficiency in conventional steelmaking can only be fully achieved by replacing old production units or expanding capacity. The steps themselves, however, rely on standard technologies which are highly cost-effective in new installations (which is one of the reasons that many new mills in developing countries are so competitive). Direct reduction processes are currently in pilot-plant development.

In terms of end-use energy, best available technologies reduce electricity use in steelmaking by 23% and fuel use by 35% (Gyftopoulos et al. 1974). The 2.5-fold increase in electric steelmaking increases electricity demand, but this can be largely compensated for by improvements in electric steelmaking (Neumann et al. 1975)—continuous casting and optimization of electric drives in conveyors, fans, compressors, and so forth (see below, "other electrical applications"). The total saving in primary energy for steelmaking—40%—is larger than the savings in end-use fuel and electricity because of the potential for more extensive cogeneration and use of byproduct gases.

Chemicals. The data in Table 2.2 show that 31% of the end-use energy to make chemicals in the FRG in 1973 was electricity, almost two-thirds of which was used in electrical processes or for process heat. The German chemical industry is highly innovative, and process technologies change rapidly. Energy efficiency can be expected to improve rapidly as old plants are replaced, the chain of intermediate process stages is shortened, and new catalytic and biological process routes are introduced (SERI 1981). Fichtner (1977) estimates that by 2000 the specific end-use fuel demand will have been reduced by 51% and specific electricity demand by 21%, based on available and cost-effective technologies. Modern power electronics and other improvements in electrical devices can boost electric drive efficiency by another third (see below). We are thus conservative in assuming that best available technologies will reduce specific consumption of both fuels and electricity in the chemical sector by 50%.

Cement. The FRG produced about 600 kg/cap of cement in 1973, using an average of 4.3 GJ/T of end-use fuel and 0.3 GJ/T of end-use electricity. This is lower than in many other countries because FRG clinker production uses the dry process. No major alternative is currently available for cement production. Kilns in the mid-1970s had an end-use conversion efficiency of 54–56% (Fichtner 1977). Heat recovery and insulation can substantially reduce losses, and more efficient grinding methods can cut electrical needs (ibid.), together reducing demand in new units to 2.8 GJ/T fuel and 0.2 GJ/T electricity—still well above the theoretical limit of 0.9 primary GJ/T (Gyftopoulos et al. 1974). We assume here these practical improvements, but ignore the considerable savings obtainable by substituting stress analysis for concrete, using less energy-intensive materials, and possibly substituting in the future stronger aggregate materials, based, for example, on autoclaved olivines (Kihlstedt 1977), whose production uses less energy at far lower temperatures (ca. 250°C).

Other electrical applications. The main use of electricity in other German industries, as in the three just described, is electric drive—for machine tools, pumps, fans, compressors, conveyors, presses, and so on. The average practical conversion efficiency of electricity into motive power at the point of use is not well known; estimates range from 40% or less (Murgatroyd & Wilkins 1976) to 60% (Schäfer 1971).[13] Improving efficiency requires ascertaining actual power requirements and sizing motors accordingly; using improved (e.g., Wanlass) motors and (where needed) torque converters or clutched flywheels; operating motors from alternating current synthesizers (ACS) or other power-factor controls; and in some cases switching to central or local (Olivier et al. 1981) hydraulic drive. Using ACS alone allows an average 30% electricity saving (Ben-Daniel & David 1979). Of these technologies, only ACS is not yet fully commercially available in all relevant sizes, though it should become so in 1981–83 on present production plans. The expected price of ACS units is $10–50/kWe of nameplate motor power, corresponding to a cost of saving electricity at least ten times lower than the capital cost of marginal nuclear electric capacity (Krause et al. 1980), so even retrofitting is economically attractive. The best electric drive technology will yield at least a 33% saving in kW-h.

Some industries outside the "big three" use electric process heat, and the "other electrical" process electricity item in Table 2.2 also includes alumina reduction—17% of the 8.8 MTCE shown. This reduction currently uses the Bayer-Hall process. The Alcoa process yields a 30% electricity saving overall (Alcoa 1973). Plasma arc reduction of alumina or of aluminum chloride, saving 40–65% compared with Bayer-Hall (Rains & Kadlec 1970), has been proposed but not yet technically developed. Some more conventional approaches (e.g., Pechiney), however,

---

13. The Schäfer estimate is based on lathes and similar machine tools.

go part of this distance and are becoming available (9–11 kWe-h/kg Al). The theoretical limit for primary aluminum production is only 13% of the 1973 FRG intensity (Gyftopoulos et al. 1974).

Conservatively, an average saving of 33% can be expected from using best available technologies in the "other electrical" industrial applications.

Other process heat applications. The average Second Law conversion efficiency of fuels into process heat (and, where practiced, into cogenerated electricity) in the FRG is <22% (Krause 1981). Available work is lost through inefficient production techniques and chemical process routes, the use of high-quality fuels for low-grade tasks, and release of uncaptured waste heat. While processes differ substantially and must be considered individually, they use only standard industrial components such as furnaces, boilers, dryers, and ovens. For these components an arsenal of energy-saving technical improvements is also available: thermal insulation; regenerators for directly fired low-temperature processes; recuperators for directly fired high-temperature processes and flue gases; waste heat boilers; combustion air preheaters; cogeneration; heat pumps; fluidized-bed heat-treating furnaces; heat cascading; and electronic process controls. Combinations of these measures permit 2–2.5-fold increases in Second Law efficiency (Thermo-Electron 1977).

These technologies are already being used in some plants, but even at 1973 energy prices they were not used as widely as was worthwhile (Fichtner 1977). Cost-effectiveness is well documented (ibid.; Thermo-Electron 1977; Gyftopoulos et al. 1974; Williams 1978; Sant 1979, 1981). Payback times at historic fuel prices are often about 5–7 years or less. But most analyses of medium-term prospects treat the introduction of such measures very cautiously, due to the often high capital costs and technical obstacles found in retrofitting industrial plants (SERI 1981). In the long term, the turnover of the old stock will obviate these problems, and cost-effective technologies can reduce specific fuel use for process heat by at least 50%.

Industrial space heat. About 85% of all industrial employees worked outside the "big three" industries, and Table 2.2 allocates the 10.7 MTCE of total industrial space heating pro rata. The same technologies and costs apply in this use as in commercial space heat, with the added possibility of re-using low-grade industrial waste heat on the premises. Available, cost-effective technologies would reduce the relative energy intensity to 0.21 with a "2030" building mix, and in the long-term technical limit, conservatively to 0.05.

Summary of industrial sector. The combined saving from available and cost-effective technologies is shown in Table 2.6 and in the "2030" bar in Figure 2.4. Industrial end-use energy to produce the FRG's 1973 output can be reduced by 45%, electricity use by 36%, and fuel use by 47%, given a reasonable 50-year turnover of capital stocks.

Table 2.6 *Cost-effective potential efficiency improvements in FRG industry*

Sector	"2030" relative energy intensity	"2030" end-use energy (MTCE)[a]		
		fuel	electricity	total
steel				21.0
fuel (including reductant)	0.65	19.05		
electric process & drive	0.77		1.93	
chemicals				7.7
fuel	0.50	5.25		
electric process & drive	0.50		2.40	
cement				3.6
fuel	0.65	3.45		
electric drive	0.67		0.20	
other industries				20.3
fuel for process heat	0.50	12.45		
fuel for space heat	0.21	1.90		
electric process & drive	0.67		5.90	
total		42.10	10.43	52.6
index (1973 = 1.00)		0.53	0.64	0.55

a. No "present technical limit" case is shown here; its estimation would involve many processes, and although considerable savings are plausible, they could easily be overshadowed by shifts in the composition of output.

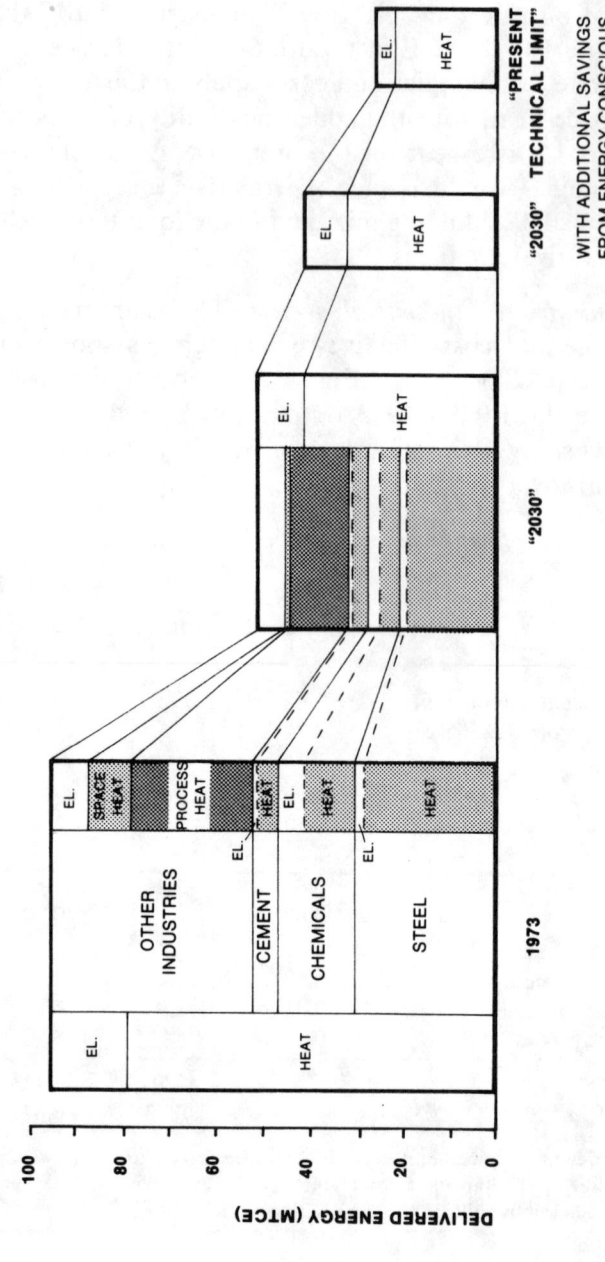

Figure 2.4 *Cost-effective potential efficiency improvements in FRG industry, at constant 1973 activity levels, including effects of materials policy*

2.3.4. Efficient use of materials

Most energy studies neglect the saving that can be achieved by using materials efficiently. Yet the saving in embodied energy—the available work that can be recovered from materials or whose commitment can be avoided by proper use of materials—can be large enough, if systematically pursued, to increase the Second Law energy efficiency of an industrial economy by severalfold (Ayers & Narkus-Kramer 1976; Krause 1981). These indirect energy savings must therefore not be ignored. They are also of special interest for economic and security reasons to countries like the FRG which import many of their raw materials.[14]

Technical options for materials efficiency fall into two classes: reducing losses from the materials cycle, and reducing excess materials in the cycle. Measures to do both include recycling materials, recycling products (by remanufacture, reworking, and re-use, depending on the stage in the process at which the measure is taken), reducing dissipative uses, reducing losses in milling and manufacturing, reducing post-consumer waste, eliminating unnecessary materials in products, and increasing product lifetimes. These measures, in short, seek to maximize (within economic and social constraints) the benefit derived from each unit of material used. Rather than saving energy by making a throw-away beer can more efficiently, or even replacing it with a reusable container, they recycle the can—indirectly avoiding energy use in industry by efficiently using industrial output, which is after all made of materials and energy.

14. This will become even more important because the energy needed to win a ton of metal rises very steeply with declining ore grade and mineral grain size (Lovins 1973; Chapman 1974), and in large steps for the many less common metals with a discontinuous grade-tonnage distribution (ibid.; NAS 1969). The eventual time when high-grade (low-entropy) resources are so depleted that the technostructure needed to recover them can no longer sustain itself can be postponed for a long time—of order 10^4 years or more (Lovins 1980)—by sensible materials policies. In the nearer term (10^2 y), mineral supply is constrained mainly by land use, water, and politics. Later, despite technological progress (Kihlstedt 1975; Kellogg 1977; Eketorp 1980), the energy needed to win each ton of most metals will rise substantially, but not enough in comparison to other societal energy needs to be important in the energy supply and demand budgets considered in this study. From now on, the substitutability of many resources (Ayres 1974; Ayres et al. 1974; Goeller & Weinberg 1975; Smith 1979) and sound materials policies will be important issues on any national policy agenda.

The technical potential of these options has been comprehensively studied for eight metals in the US (OTA 1979, 1979a). Recovery of primary metals from post-consumer solid waste has been investigated in both the US and the FRG (OTA 1979; Turowski 1977). The waste plastics, paper, glass, ferrous metals, and tires available in the FRG have been quantified (Umweltbundesamt 1979). For nonferrous metals, the FRG substitution and recycling potential was studied by Studieck et al. (1976). The energy-weighted recycling fraction of industrial material output in the FRG in 1974 was only about 11% (Krause 1981).

As shown in Table 2.7, about 9% of energy use in the listed industries can be saved by eliminating excess material, and another 7% of all industrial energy use by expanding the fraction of secondary material to levels practically achievable without undue inconvenience (Krause 1981). Even these conservatively expanded levels of material and product recycling (extrapolating the former also to sectors not specifically shown in Table 2.7) decrease total industrial energy use by 22%. (This is *not* the same as a shift in composition of output or a shift from energy-intensive towards skill-intensive industries or a shift from industry to services; our analysis assumes none of these changes.) With a more systematic and sophisticated approach to re-use and a shift towards recycling of products rather than of materials, the effectiveness of recovering embodied energy could well be doubled from 10% to 20% in the long term (ibid.), reducing industrial energy demand by about 50%—our "technical limit" assumption.

Recycling systems and technologies are for the most part well-known and in commercial use today. Under present conditions, the recycling fractions in Table 2.7 may exceed what can be sold at an attractive price, but economic conditions for recycling are rapidly improving as energy, virgin ores, and waste disposal become costlier. Scrap price variations with the business cycle, and other problems, can be overcome by proven institutional means (OTA 1979).

Table 2.7 *Potential energy efficiency improvements from reducing excess and lost materials and from increasing recycling of materials*

Material	(1) % of industrial end-use energy	(2) Energy saving from excess materials	Secondary material fraction (3) Increase from	(4) Specific to energy saving		$(1)\times(\Delta 3)\times(4)$
steel	0.33	15%	0.4	0.6	55%	0.0363
cement	0.06	15%	0.0	0.0		
thermoplastics	0.05	20%	0.02	0.50	96%	0.0230
paper products	0.035	10%	0.33	0.50	40%	0.0024
glass	0.025	24%	0.15	0.45	20%	0.0015
fertilizer	0.01	50%	0.0	0.0		
aluminum	0.016	15%	0.25	0.50	68%	0.0027
weighted total	0.526	8.54%				0.0659

combined saving: 22% of industrial end-use energy

SOURCE: Krause 1981.

Summary of efficient use of materials. The combined energy saving from recycling the materials listed in Table 2.7 would be 6.6% of the total 1973 industrial energy end use, as shown in the last column. Reducing excess materials in the cycle—column (2)—would save 9% of the energy in those industrial sectors, or, if generalized to all industrial output, 0.0854/0.526 = 16.2% of all industrial energy. The combined energy saving in all German industry would then be 22%. There would be a significant additional saving in fossil fuels used as petrochemical feedstocks (Section 2.3.5). Neglecting that feedstock saving for the moment, Table 2.8, summarized in the last two bars of Figure 2.4 above, shows the combined effect of direct and indirect (materials) energy efficiency on 1973 FRG industry.

The additional effect of including materials savings is thus a factor of 1.28 initially and a further factor of 1.56 ultimately, generously assuming no further "technical limit" improvement in direct industrial energy efficiency.

Table 2.8 *Technical potential of combined efficiency improvements in direct and indirect FRG industrial energy use*

	Relative energy intensity (1973 = 1.00)	
Mode of saving	"2030"	"present technical limit"
direct	0.55	(no improvement assumed)
indirect	0.78	0.50
total	0.43	0.275

2.3.5. END-USE AND PRIMARY ENERGY: THE POTENTIAL OF EFFICIENT CONVERSION

Table 2.9 summarizes the 69% end-use energy saving that can be realized with best available and cost-effective technology; 79% in the longer term. These two cases are shown by the solid bars in Figure 2.5.

Table 2.9 *Cost-effective potential efficiency improvements in energy end-use in the FRG*

	Relative energy intensity (1973 = 1.00)[a]		End-use energy (MTCE)			
			"2030"		"technical limit"	
Sector	"2030"	"technical limit"	fuel	electricity	fuel	electricity
residential & commercial	0.18	0.10	16.0	3.9	7.5	3.7
transport	0.36	0.32	17.2	0.8	15.1	0.8
industrial & materials	0.43	0.27	32.9	8.2	21.1	5.2
total			66.1	12.9	43.7	9.7
total of fuel plus electricity			79.0		53.4	
index (1973 = 1.00)			0.30	0.42	0.20	0.32
index of fuel plus electricity (1973 = 1.00)			0.31		0.21	
end-use structure: fuel for heat (1973 = 69%)			48.9 (62%)		28.6 (54%)	
fuel for transport (19%)			17.2 (22%)		15.1 (28%)	
supplied as electricity (12%)			12.9 (16%)		9.7 (18%)	

a. Composite indices for fuel and electricity, shown for comparison only; the last four columns are calculated using the separate indices for fuel and for electricity, as shown in Tables 2.4, 2.5, and 2.6.

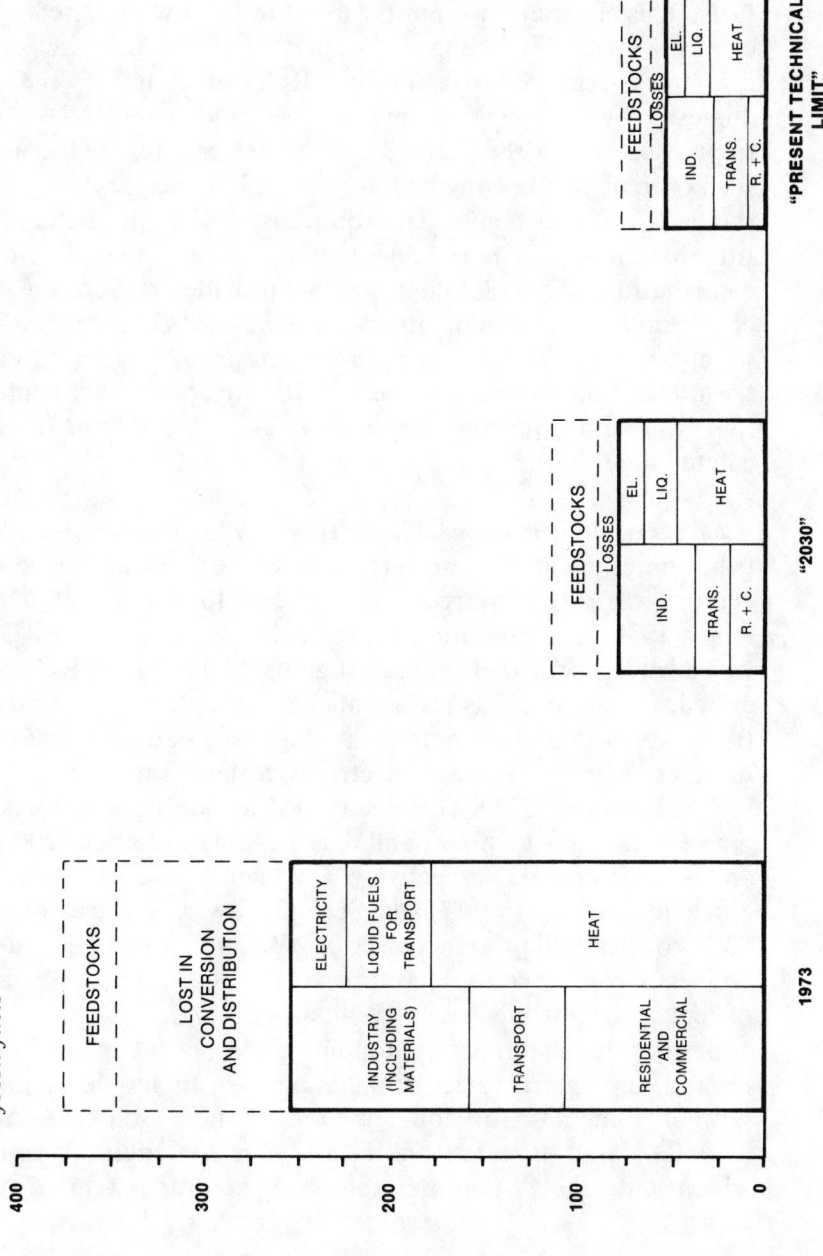

Figure 2.5 *Technical potential of cost-effective efficiency improvements in FRG end-use and primary demands at constant 1973 activity levels, assuming all future end-use energy is converted from fossil fuels*

In order to derive the corresponding primary fuel demand, we shall assume here that all end-use energy is derived from fossil fuels. This calculation is not meant to reflect the expected mix of energy systems (including renewables) that would in fact supply primary energy to an efficient FRG: for example, the 1973 domestic hydroelectric output alone would meet 17% of electricity demand in the "2030" case. Later chapters will consider such options for meeting end-use needs. For the present, we seek rather to offer a convenient comparison with the actual 1973 situation, in which renewable sources made a negligible direct contribution. (The solar energy which provided most of the space conditioning—it would otherwise be -237°C outdoors—and which grew the food and fiber crops was probably larger than all the direct energy use, however.) With any renewable contribution, the actual primary demand would be less than that calculated here.

Conversion efficiency. The average 1973 conversion and distribution efficiency of primary fuels into energy at the point of end use was 87% for direct fuels and 31% for electricity (Meyer-Abich 1978). Cogeneration technologies, in which the FRG is a world leader, provide the greatest gains. In 1973 the FRG cogenerated, mainly in industrial plants, an estimated 13% of all electricity generated. The main technology was backpressure steam turbines, with an average electricity-to-heat ratio of 50 kWe-h/GJ (Fichtner 1977). Other standard technologies are gas turbines, steam/gas turbines, and diesel engines, at about 200, 320, and 400 kWe-h/GJ respectively. Cogeneration offers First Law efficiencies of up to 90% and Second Law efficiencies of up to 50%, corresponding to increases of 25% and 46% in Second Law efficiency compared to separate production of process steam and of central-station electricity (Williams 1978).

Besides these commercial topping cycles, heat engines for bottoming and intermediate applications are under development: Nitinol™ engines for low-temperature heat sources (-50 to +60°C), small organic Rankine and Ericsson engines—some of which are now commercial—for medium temperatures

(65–550°C), and high-temperature Stirling engines (650–800°C) (*Soft Energy Notes* 1980a).

Over the years, the ratio of industrial electricity demand to process heat demand in the FRG has increased to an average of about 60 kWe-h/GJ, with values >200 in some industries, whereas older steam-based cogeneration units deliver only 50. Better matching with modern steam- or gas-turbine and diesel cogeneration systems has been inhibited by unfavorable utility buyback prices (Meyer-Abich 1978). Industrial managers must also note that conservation technologies which reduce steam requirement compete with, and are generally more energy- and money-efficient than, cogeneration of low-temperature process steam.

Mix of conversion technologies. Based on the 1973 FRG temperature spectrum of process heat supplies (Fichtner 1977) and on the end-use efficiency improvements discussed in Section 2.3.3, about 12 MTCE of low-temperature process heat can be cogenerated in the "2030" case and about 7 MTCE in the "technical limit" case, all cost-effectively against marginal and (usually) present prices. If replacement of old units raises the average cogeneration output ratio to 200 kWe-h/GJ, 70 and 47 TWe-h/y can be cogenerated respectively, compared to 30 TWe-h in 1973. These outputs correspond respectively to 67% and 89% of all electricity needs in the entire economy after efficiency improvements.

In the hypothetical case of cogenerating electricity only from fossil fuels, assuming average First Law efficiencies of 75% for cogeneration and 90% for converting primary to delivered direct fuels, the required primary fuels to supply the FRG would be 90.6 MTCE in the "2030" case and 61.5 MTCE in the "technical limit" case, as shown in Table 2.10 and by the dashed lines in Figure 2.5 above. While many heat applications in industry and hot water uses in buildings allow for baseload generation, there are some seasonal and business cycle variations that would need to be met by means not tied to industrial heat needs. In a system using only fossil fuels, this compensation would reduce the average First Law efficiency of electrical generation somewhat below that of cogeneration. In reality, of course, at least the hydro capacity available in 1973 would also contribute to baseload generation, and the potential of wind and of direct solar electricity (thermal or photovoltaic) is far larger. To approximate these competing effects, our calculation of primary fuel requirements assumes the lower end (75%) of the range of First Law efficiencies for cogeneration.

Backpressure turbines, gas turbines, and diesel and Otto engines in sizes ranging from 15 kWe (TOTEM) to 1 MWe (neighborhood diesel [Hein 1979]) and 30 MWe (large industrial turbines) are commercial cogeneration technologies in the FRG, where several small municipal utilities already operate gas-fired local diesel total energy systems. These systems have an installed price of about $500/kWe and successfully compete against present average fuel and electricity prices (ibid.). The TOTEM total energy units cost $430–500/kWe installed and are also cost-effective at present FRG prices (*Soft Energy Notes* 1980a). Backpressure steam turbines of about 30 MWe are cost-effective against new coal and nuclear central power plants (Williams 1978; Fichtner 1977), delivering electricity at $0.035/kW-h (@ DM 1 = $0.5) as compared with $0.055 from central stations (Fichtner 1977). (We neglect here an option that may provide major electricity savings in some industries, notably steel: rather than cogenerating in order to run electric drives in the plant, it is often cheaper to eliminate the electric middleman and drive the motors directly with gas or steam turbines [Olivier et al. 1981].) The lower-temperature heat engines mentioned above, and small TOTEM units with outputs of only 10 kWt, are generally in prototype stage, although some manufacturers, such as Hormat Turbines in Israel, are selling Rankine systems in the 75–95°C range for solar pond applications.

Feedstocks. Of the 29.9 MTCE of fossil feedstocks used in 1973 (Table 2.1), 7% was coal, 83% oil and naphtha, and 10% natural gas. The main products were plastics, fertilizer, and other chemicals. We assume here that the savings of 21% ("2030") and 50% ("technical limit") derived earlier for increased materials efficiency can be applied also to feedstock requirements. This saving is reflected in Figure 2.5.

Summary of all sectors. Converting the foregoing results to primary energy terms, as shown in Table 2.10 and Figure 2.5 above, yields a total primary energy use (including feedstocks) of 1.7 kW/cap ("2030") or even as low as 1.1 kW/cap ("present technical limit"). It is important to note the many conservatisms in both these figures. They assume that efficiency improvements stop far short—often an order of magnitude short—of the price

Table 2.10 *Technical potential of cost-effective efficiency improvements in FRG primary energy use*

	Primary to end-use conversion efficiency		Primary energy demand (MTCE/y)	
Application	1973	"best"	"2030"	"present technical limit"
fuels (transport & heat)	0.87	0.90[a]	73.4	48.6
fuels for electricity	0.31	0.75[ac]	17.2[c]	12.9[c]
total fuels	0.73	0.87	90.6	61.5
index (1973 = 1.00)			0.26	0.18
feedstocks			23.9	15.0
index (1973 = 1.00)			0.80	0.50
total fuels & feedstocks			114.5	76.5
index (1973 = 1.00)			0.30	0.20
per capita[b] kW: end-use energy	(1973 = 3.80)		1.18	0.80
+ system losses	(1973 = 1.42)		0.18[c]	0.12[c]
= primary fuel	(1973 = 5.22)		1.36[c]	0.92[c]
+ feedstocks	(1973 = 0.45)		0.36	0.22
= primary energy	(1973 = 5.67)		1.72[c]	1.15[c]

1 MTCE = 29.3 PJ = 27.8 × 10^{12} BTU = 8.14 TWt-h = 5.05 million bbl
    = 7 × 10^{12} kcal
1 MTCE/y = 9.285 × 10^8 W = 0.01498 kW/cap[b]

a. These values are used to derive the demands in the last two columns from the fuel and electricity totals in Table 2.9: e.g., for "2030" direct fuels, 66.1 ÷ 0.90 = 73.4. The 1973 and "best" "total fuels" efficiencies are for comparison.
b. Using the 1973 FRG population of 61.98 million for all columns.
c. Assuming conversion entirely from fossil fuels with the efficiency described above.

already being paid for marginal supplies, and that feedstock demand is not very sensitive to price. They assume no significant changes in lifestyle, no new technical discoveries (which are in fact announced almost daily), no measures to accelerate implementation, no shift of societal values from the acquisition of ever more consumer ephemerals and the subordination of people to machines. And they assume no redefinition of tasks: what Second Law efficiency measures, and what our "technical fixes" consider, is only the minimum energy needed to provide a given change of state. But if, for example, information can be moved rather than bodies, or low-energy materials or stress analysis substituted (as mentioned earlier) for Portland cement, then the final state—the task to be done—has been redefined and the desired human end (communication or building) can be met with even less energy than we have supposed.

2.3.6. Findings and international comparisons

The analysis in the previous five sections has shown how, through technical measures that stop well short of what is technically feasible or economically worthwhile, to increase by a factor of 3.3–5.0 the total primary energy productivity of the 1973 economy of the Federal Republic of Germany—already one of the most energy-efficient and heavily industrialized countries in the world. Stricter application of the criterion of minimizing marginal costs, or more aggressive measures to implement measures faster than by the leisurely workings of a market that too often perceives only average costs, or further technological progress, or changes in societal preferences and lifestyles, could each considerably raise the savings.

This conclusion is not unique to Germany. It is consistent with a rapidly growing body of even more detailed analyses from other industrialized nations. Among the first, and still among the best, of these (Leach et al. 1979), sponsored by the Ford Foundation during 1977–78, showed how the systematic use of "technical fixes" (generally cheaper than 1976 North Sea gas at about \$3.5/GJ) in over 400 sectors of the British economy could treble UK primary energy efficiency—about the same as our result for

Germany, which starts with a more efficient economy but competes with somewhat higher prices. In particular, Leach et al. (1979) showed that if, from 1976 to 2025, real UK GDP increased to 3.1 times its initial level, industrial production to 2.2×, car travel to 1.8×, and air travel to 3×, nonetheless total primary energy use could decline to 0.93× its 1976 level. The assumed rate of implementation of these energy-saving measures was considerably *slower* than had actually been exhibited from 1974 to 1977, and slower still than the rate from 1978 to 1980. Exhaustive review of these findings by British energy agencies and industries and by international bodies (including IIASA) failed to refute them: the study was, as claimed, technically conservative (especially in the industrial sector), often more so than UK government forecasts, and applied the weak economic test of near-1976 prices for the cheapest domestic fuel. The aim of the Leach study, like ours in this paper, was to say what could, not what will, happen: to be explicitly rather than tacitly normative. Like our analysis, it began with a target date far enough in the future that major capital stocks could be largely modified, replaced or diluted—and then, as we shall do below, worked backwards towards the present to see what must be done when.

At first privately and then under UK government auspices, D. Olivier et al. (1981) have carefully enlarged upon Leach's pioneering work. Olivier originally intended to see how far Leach's trebled GDP/E ratio could be bettered by assuming technical fixes which might not be cheaper than \$3–4/GJ but were at least cheaper than new synfuel or power plants (assuming prices for these well below official estimates). Though Olivier's technical assumptions go further than Leach's, they are also conservative. They are defensible in detail as realistic and not pressing any technical or political limits. The Olivier analysis is indeed the most detailed efficiency scenario so far done anywhere in the world: his treatment of the steel industry alone, for example, is longer and more elaborate than this summary for the entire FRG economy. In large part because of this disaggregation, his explicit list of documented technical fixes nearly doubles Leach's E/GDP improvement. Although real GNP is assumed to increase to 2.9× its 1976 value, and industrial production to 2.3×, total UK primary energy use nonetheless drops to 0.55× its 1976 value—an overall primary energy reduction of 5.27× per unit of real GDP. Yet, as Olivier then found to his chagrin, his main energy-saving measures were *still* roughly competitive with North Sea gas prices, and far below marginal prices for long-run replacements. Optimized against marginal price, the achievable UK energy saving—not yet assessed in detail—is probably at least eightfold.

Many of Olivier's technical assumptions are similar, especially in the industrial sector, to those of Krause (1980), who found in the more efficient FRG economy that the E/GDP ratio could be reduced by a factor of 3.6 to 3.85, depending on whether projected structural shifts in the composition of output of the industrial sector were taken into account. J.S. Nørgård and colleagues at the Technical University of Denmark have likewise shown in an increasingly detailed series of books (Nørgård 1979) and papers (Meyer et al. 1977–79) how to more than double Danish primary energy efficiency. Their work is notable for its original treatment of the dynamics of implementation. Equally conservative in its technical assumptions—even more so than Leach et al.—is an analysis (MALTE 1977; see also Johansson & Steen 1978) of how to decrease 1990 primary energy use in Sweden—already arguably the most energy-efficient industrial nation (Schipper & Lichtenberg 1976)—to 0.87× its 1976 level despite assumed increases of 1.45× in real GNP and 1.51× in industrial production: a 40% drop in energy/GNP ratio. (The Swedish Government's 1978 Energy Bill proposed a 32–38% drop by 1990 [Foley 1979].) More recent Swedish analyses show 2–3 × savings over 25–30 years (Steen et al. 1981).

A distinguished panel of the US National Academy of Sciences CONAES study has published (CONAES 1978) American projections for a doubled real GNP in 2010. Despite strong technical and economic conservatisms—for example, assuming in all but the lowest scenario that the average US house built in 2010 will be less heat-tight than the average Swedish house is today (Schipper & Lichtenberg 1976)—and despite extensive compromises required for consensus among a diverse group, the Panel projected "pure technical fix" changes in

E/GNP to 0.49–0.62× the 1978 level, depending on the vigor of policy actions, or to 0.40× with modest changes in lifestyle. (Some members believed this could be achieved without those changes by using more realistic technical assumptions—a belief buttressed by comparison [USDOE 1980d; Schipper & Tuininga 1979] with Leach's technical coefficients, which are generally comparable with the highest of the three US scenarios.) A study for the US Department of Energy (Craig et al. 1978) analogously showed how California's economic activity could treble by 2025 with only a 20% increase in end-use energy needs. Making the technical assumptions more realistic and internally consistent (Lovins 1978a) could readily turn this into a 22% decrease (decreasing E/GNP to 0.26× its initial level).

More recently, three important analyses for the US, of global significance in their detail and methodology, have strongly confirmed these results. The first two examined what combination of investments from 1968 to 1978 (Sant 1979) or from 1980 to 2000 (Sant 1981) would supply desired energy services at least direct cost to the consumer—that is, optimizing against "rolled-in prices" (which average high marginal with low historic costs). The 1968–78 simulation found that economically efficient investment would have reduced 1978 US purchases of oil, coal, and electricity by about 28%, 34%, and 43% respectively, while simultaneously reducing by about 17% the price of providing the same energy services that actually were provided. Suspecting that recent price increases would elicit an even stronger efficiency gain, Sant then prospectively simulated the next two decades in even greater detail (his model contains many hundreds of technologies)—but still assuming rolled-in prices based on official 1979 forecasts, which by 1981 were already being overtaken by actual prices many years ahead of schedule. Despite these rather weak economic tests, Sant et al. found (1981) that even 20 years of pure market competition would be enough to halve US primary energy needs per unit of GNP: GNP growth by 72% would raise primary energy use by only 9%. Investment in conventional supply, especially power stations, would essentially cease because such options cannot compete with efficiency improvements. Imported oil would decline to nearly zero because it is (aside from synfuels and power plants) the highest-cost competitor: it has already priced itself out of the market. Most interestingly, the fraction of GNP devoted to buying energy services would go *down* with efficient investment, not up. The energy sector, far from driving inflation, would become a net exporter of capital to the rest of the economy!

An equally detailed and original independent study by dozens of consultants for the Solar Energy Research Institute, commissioned by Deputy Secretary of Energy John Sawhill in 1979, also considered a US "least-cost energy strategy" for 1980–2000 (SERI 1981). The draft report, rich with revealing data, shows a larger energy saving than Sant (1981): despite an 80% increase in 1977 real GNP by 2000 and comparable increases in personal comfort and mobility, primary energy use would *decrease* to 13–18% below the 1980 level. The use of nonrenewable fuels would drop by nearly half, and total demand for electricity would probably decline. These results are quite consistent with Sant's. SERI's assumed energy prices, though still well below realistic marginal costs, were somewhat above those which Sant assumed, eliciting larger savings. Yet SERI's technical assumptions, like Sant's, were conservative: every technology considered for 2000 has already been demonstrated in the US.

Less detailed and even more conservative, but still persuasive, efficiency scenarios have been published for Canada, both federal and provincial (Brooks 1981, 1981a), Switzerland (Peter 1977; Schweizerische Energiestiftung 1978; Ledergerber 1979), France (Groupe de Bellevue 1978; Les Amis de la Terre 1978; Giry 1978; Bosquet 1978); Japan (Institute of Energy Economics 1980); and elsewhere. There are also many other studies for countries already mentioned, for example by Ross & Williams (1981) for the US. Many of these have uniquely useful features. Taken as a whole, this literature proclaims a clear message: *every* industrial country can cost-effectively improve its energy efficiency by severalfold, using only presently available technical measures.

The gradual penetration of this insight into official planning circles has led to a remarkably consistent downward trend in most countries' energy demand forecasts (e.g., Marshall 1980;

Norman 1981). The US example shown in Table 2.11 is a diagonal matrix with a two-year time constant for transitions towards the lower right in a direction of increasing sociological respectability. (Publishing it in 1978 unfortunately made people aware of this process, so that some of the 1980 forecasts—notably the Sawhill Report [SERI 1981]—tend to skip a notch, making the proper column classification of some of the 1980 entries somewhat debatable.) The decline has nowhere near hit bottom yet: for example, it does not reflect the relatively recent insight (Ayres & Narkus-Kramer 1976) that an energy-conscious materials policy could roughly treble US Second Law efficiency. Analysis of this option and use of more realistic technical coefficients on the lines of Krause (1981) and Olivier et al. (1981) respectively would arguably lead, in a few more rows of the matrix, to long-run (post-2050) US primary demands around

Table 2.11 *Evolution of approximate estimates of US primary energy demand in the year 2000 ($q/y \equiv 10^{15}$ BTU/$y \cong EJ/y \cong$ 36 MTCE/y (1980: 76 q/y) $\cong$ 2.54 TW)*

Year of forecast	Source of forecast			
	Beyond the pale	Heresy	Conventional wisdom	Superstition
1972	125[a]	140[b]	160†[c]	190†[d]
1974	100[e]	124[f]	140†[g]	160†[h]
1976	75[a]	89[i]–95[j]	124[g]	140†[h]
1978	33†[k]–55[a]	63†[m]–77[m]	95[n]–96[m]–101[p]	123[q]–124[r]
1980*	15[a]	49†[s]–54†[t]	62[u]–64[v]–66[w]–84[x]	97[y]–100[z]

a. Lovins speeches.
b. Sierra Club.
c. US Atomic Energy Commission.
d. Other Federal agencies.
e. Energy Policy Project "Zero Energy Growth."
f. Energy Policy Project "Technical Fix."
g. US Energy R&D Admin. (ERDA).
h. Edison Electric Institute.
i. Von Hippel & Williams (Princeton).
j. Lovins, *Foreign Affairs*, IX.76.
k. Steinhart (U. Wisc.), 2050.
m. CONAES (1978) I-III, 2010.
n. USDOE DPR, IX.79, $32/bbl (1977 $).
p. Weinberg (IEA-Oak Ridge) "Low."
q. USDOE DPR, IX.79, $18&25/bbl av.
r. Lapp (believes E/GNP fixed).
s. Stanford III, 2010 (USDOE 1980d).
t. CONAES (1980) CLOP, 2010.
u. SERI (1981) "Sawhill Report" low.
v. Ross & Williams (1981).
w. SERI (1981) "Sawhill Report" high.
x. Sant (1981) "least-cost."
y. Exxon *World Energy Outlook*, XII.80.
z. USDOE NEP-3 midrange.

*Formal publication of several of these entries was delayed until early 1981, but all were available to scholars during 1980.
†With lifestyle changes.

10–15 q/y—four to six times below present US energy use—even if real GNP approximately doubled (see Table 2.17 below).

As efficiency scenarios have become technically more disaggregated, they have also—equally importantly—become geographically more disaggregated. In the US in 1980–81, for example, probably over a thousand efficiency scenarios were being studied (or implemented) at a state, county, city, or town level. Specialized methods and tools for this purpose are becoming available (SERI 1980). It is often turning out in industrialized countries, as in India (Reddy 1980), that localized analysis "from the bottom up" shows the greatest savings.

Efficiency scenarios are still commonly misinterpreted, despite explicit disclaimers, as assuming or entailing drastic social change for the worse (Nash 1979). The studies described here offer, on the contrary, a means of *avoiding* such change through the intelligent application of proven technologies guided by the principles of market economics. They assume no significant change in where we live, how we live, how we run our societies—save those changes unavoidably resulting from a doubling or trebling of GNP (an assumption which would lead to a per capita National Product grosser than anyone else's, and which certainly exercises the imagination). Many analysts of efficiency scenarios, including ourselves, consider such undifferentiated economic growth, in M. Goldberger's phrase, "spherically senseless"—it makes no sense no matter which way round one looks at it—but we have assumed it anyhow to save argument, keeping our personal preferences separate from our analytic assumptions. In practical effect, we argue that if one wishes to Los Angelize the planet, efficiency scenarios can fuel that process most cheaply: indeed, that underinvestment in energy productivity and overinvestment in costlier energy supply would seriously *inhibit* economic growth.

If enough people felt that a goal of mere swelling was unworthy, or considered today's values and institutions somehow imperfect, the resulting social change would merely create larger energy savings than we have shown. Conversely, if economic activity (or population) should grow *faster* than assumed in the above studies, then energy-inefficient capital stocks could be diluted or replaced even faster, so energy needs would not grow proportionately more quickly. In fact, most of the national analyses described above, including for example SERI's (1981) and Sant's (1981) for the US, assume economic growth rates far above those typical of the past decade—often twice as large. The conclusion is inescapable: that if we follow the principles of economic efficiency to which the market economies (and, in different terms, the centrally planned economies too) profess allegiance, improved energy productivity should lead the industrialized countries to expect gradual energy *shrinkage* to the year 2000 and beyond, even if their most sanguine economic targets are reached. (Rates of implementation for energy savings will be discussed in Chapter 5.)

2.4. Regional and global implications

Exploring the effects of economically efficient energy investments on the fossil-fuel burn requires assessment of total energy demand (and, in Chapter 4, of alternative sources) on a global scale. Extending national results, such as those described in Section 2.3, to the world can be done by analogy, scoping calculation, or explicit aggregation of main terms. We shall use a combination of these methods, striving to keep the calculations simple and transparent.

For initial illustration, consider the United Nations Economic Commission for Europe (ECE) area, comprising North America, Western and Eastern Europe, and the USSR: 1.0 billion people using three-fourths of global energy. An earlier study (Lovins 1978) combined reasonable population and economic projections for this affluent region (which includes all major industrial countries except Japan) with efficiency assumptions more conservative than those demonstrated by the foregoing national analyses. Nonetheless, if by 2025 in ECE:

- population grows to 1.25 billion (in line with official projections);
- gross economic activity per capita increases 82%;

- the output ratio of goods to services (with an assumed three-fold difference in average energy intensity) decreases 40% in line with current trends;
- the ratio of primary to end-use energy drops from 1.35 to 1.05 through careful thermodynamic matching, total-energy systems, and renewables; and
- technical end-use efficiency improves by average factors of 4.0 in North America and 2.5 in Europe and the USSR—considerably less than suggested by the technical coefficients documented earlier;

then the resulting primary (end-use) energy needs of the ECE region in 2025 will be about 107 (102) EJ/y (1 EJ ≡ 10^{18} J) or 3.4 (3.2) TW-y/y—40% *less* than in 1975, despite the assumed 2.28-fold increase in economic activity.

This encouraging result does not yet take account of the aspirations of the other 3.4 billion people in the world. High-energy forecasts commonly assume that developing countries must and will imitate, indeed recapitulate, the pattern of development which served the OECD nations in a very different period, when OECD enjoyed ready access (via colonial control) to cheap raw materials at competitively depressed prices and sold the resulting manufactures at monopoly rents. Today, with conditions rapidly reversing, it is hard to see how such a pattern could be reproduced. The emphasis of development economics is therefore shifting from urban industry—based on capital-intensive high technologies, supplied by OECD nations and hoped to produce "trickle-down" rural benefits—to a more indigenous rural development based on sustainable agriculture, capital-saving and equity-fostering technologies, and meeting basic human needs. Would this pattern, on which there is a consensus among the international development agencies, require vastly more energy?

Developing countries tend to have poor energy statistics (IEA 1978) which generally omit their noncommercial fuels—probably about 1 TW in global total, but possibly as high as 2 TW. Much recent growth in commercial fuel use is mere substitution for dung and for about 1 T/cap-y of firewood (Eckholm 1976).

(About half the world timber cut is burned by 80–90% of people in the poorer nations.) Analyses of end-use patterns are sparse (*Soft Energy Notes* 1978–79; Goldemberg 1978; Reddy 1980), but representative Indian village data, qualitatively characteristic for over 2 billion people (Goldemberg 1978), are approximately: 218 W/cap total domestic sector (96% for cooking); 76 W/cap agriculture (mainly pumping and plowing); 23 W/cap manufacturing; 15 lighting; 11 transport; total 344 W/cap, supplied as 226 noncommercial fuels, 48 animal and 33 human work, 26 oil (mainly kerosene), 5 coal, and 6 electricity. Careful analyses (ibid.; Hayes 1977; Makhijani & Poole 1975; Makhijani 1976; Leach 1979; Reddy 1978, 1980; Reddy & Prasad 1977; *Soft Energy Notes* 1979) have consistently shown great scope for improved technical efficiency in these functions, starting with innovations as simple as replacing open cooking fires (5–10% efficient) with cheap clay stoves (30–60% efficient) (*Soft Energy Notes* 1979). Such investments in efficiency are generally agreed to be the best immediate buy. Without them, neither OECD-style energy supply nor the eco-development alternatives with their many attractions—biogas, solar heat, wind, microhydro, pyrolysis—can long succeed (Howe 1977; Brown & Howe 1978; Leach 1979; Makhijani & Poole 1975; Reddy & Prasad 1977; Reddy 1980).

What is at stake for development is neatly encapsulated in Reddy's comparison, using nominal mid-1970s empirical data from India, of two ways to make nitrogen fertilizer, as shown in Table 2.12.

Table 2.12 *Two ways to make 200,000 metric tons of fixed nitrogen per year*

Characteristic	Western-style coal-fed fertilizer plant	Indian-style gobar gas plant[a]
number of plants	1	26,450
capital cost ($ million)	140	125 (ca. 60 with Chinese technology)
foreign exchange ($ million)	-70	0
direct employment (thousands)	1	131
energy balance (TWt-h/y)	-0.1	+6.35

a. Reddy 1978, 1980.

The gas plants, if sensitively embedded in the local institutional structure to ensure equity (Makhijani 1976), offer the further integrative advantage found by Reddy (1980). Currently, the cowdung is burned for cooking, so all of the nitrogen and most of the heat goes up in smoke (which blinds people). The biogas plants offer an improved fertilizer (*Soft Energy Notes* 1978a, 1979a, 1981a), a cleanly and efficiently usable fuel (which, with Chinese designs, is under sufficient pressure to pipe around the village), a substitute for dwindling firewood, and a source of efficient lighting and of running pumpset engines. The gas output can about cover all the cooking, lighting, and pumping needs of the village—about half of India's total energy needs today (Reddy 1980). Similar incentives have already led China to install about 9 million anaerobic digesters—at least 1–2 million of which are reportedly biogas producers—since 1972, half of them in a three-year period.

Some analysts, however, especially in the most industrialized countries, propose that every developing country should and will in the fullness of time become an imitation FRG, and that this will require enormous amounts of energy. Let us assume that such heavy industrialization—in addition to, not as mere displacement or migration of, present OECD industries—occurs worldwide. Yet the same ultimate efficiency levels achievable in the FRG should also be achievable by its imitators, *but much faster*. The developing countries can do things right the first time. Building anew rather than retrofitting, they can become energy-efficient faster and cheaper. They are not yet burdened by OECD's $20-odd trillion worth of energy-inefficient capital stocks with turnover times exceeding 20 years. Rapid population and economic growth would both mean *faster* accumulation of highly efficient energy-using devices to dilute the old, hence faster achievement of a highly competitive trading position. That is why some of the world's most efficient steel mills are in developing countries, demonstrating levels of energy productivity that the average OECD steel mill will take decades to achieve.

Whether developing countries will buy truly modern (or modernized ancient [Butti & Perlin 1980]) passive solar buildings rather than sealed glass monuments to declining-block electricity tariffs, best technologies rather than outmoded cast-offs, depends on their originality and shrewdness in buying. (IIASA patronizingly assumes [Basile 1981] that they "lack sufficient sophistication.") It also depends on the example set by the OECD countries. Rich, technically skilled countries with indigenous fuels, who treat energy efficiency and renewable sources as second-class and unworthy of prestige, must not be surprised if countries lacking those advantages come to similar conclusions. Industrialized countries' counsel to developing countries would be politically and morally more palatable if they first got their own houses in order.

Many observers who sympathize with the goal of world equity implied by forecasts of world industrialization (if not by Kahnian "two billion Chinese driving Buicks" scenarios) nonetheless believe that reproducing resource-intensive, pre-oil-shock OECD lifestyles on a world scale is improbable for reasons other than energy, and that achieving such distributional equity, if it can be done, will require a huge transfer of wealth from rich to poor countries (another opportunity for embodying the best energy-saving technologies). This argument only implies that energy is not as near or unavoidable a constraint on industrialism as water, soil fertility, ecological and social resilience, war, and so forth (President's Council on Environmental Quality and Department of State 1980; Lovins 1976). If for argument's sake, however, we ignore these more proximate constraints, the energy needs of a heavily industrialized but economically efficient world can be estimated from European "existence proofs."

For example, the foregoing FRG analysis suggests that if in the course of achieving worldwide the 1973 FRG level of material affluence—undreamt-of wealth to most of the world's people—the present, highly cost-effective technical limit of efficiency were reached, the primary fuel budget would average only 0.92 kW/cap. Thus, in round numbers, a 1-kW/cap primary fuel budget could efficiently support the present Western European material standard—an energy level equivalent in traditional (if repugnant) units to the food input of nearly seven 3000-kcal/day slaves per capita. For an asymptotic world population of 8 billion (again ignoring other constraints [ibid.]), this implies a rate of world energy use of 8 TW (8615 MTCE/y). That is slightly below today's rate, and far below other estimates that have rightly caused climatic concern, such as IIASA's 22–36 TW (Rogner & Sassin 1980), formerly about 30–50+ TW; Rotty's 32 TW (1980); Niehaus's 50 TW (1980); and even Colombo's 16 TW (Colombo & Bernadini 1980). The difference is simply energy efficiency. Further, given that efficient use, then even the most restrictive assessments agree that, on average, total global energy needs can be supplied entirely by renewable sources (IIASA 1981; Caputo 1980). (The additional 1.6 TW of feedstocks could be met from biomass wastes with more efficient conversion processes or with ecologically sustainable methods of extended biomass production now under development [*Soft Energy Notes* 1981; Ferchack & Pye 1981, 1981a].) Alternatively, fossil fuels could be used at a fifth of the 1975 rate of consumption—stretching the oil and gas resources alone to several centuries' duration.

Moreover, this 8-TW *Gedankenexperiment* is conservative in many respects. It caricatures an outmoded, early-1960s development concept. Its efficiency assumptions are far short of the levels economically worthwhile against the marginal cost of supply. It assumes no technological progress or changes in lifestyles or values. It assumes, obviously not correctly, that everyone lives in a climate as cold as Germany or Britain (19% of the FRG primary demand shown in the "2030" case in Table 2.10, and 8% in the "technical limit" case, is for space heating, so without these the 1.37 and 0.92 kW/cap primary fuel demands would become respectively 1.11 and 0.85 kW/cap). It assumes in the "2030" case capital stock turnover rates which are among the slowest in the world, whereas the far more dynamic growth assumed in the developing countries would produce much faster turnover. It assumes the conversion and distribution losses associated with an entirely fossil-fuel-based, relatively centralized energy system, thus incurring losses of 0.18 (0.12) kW/cap which an appropriate renewable supply system would virtually eliminate. And it assumes highly optimistic levels of gross economic activity, industrialization, and distributional equity for a doubled world population.

There is thus evidence to suggest that humankind could, in principle, avoid the potential climatic impact of prolonged fossil fuel consumption (assuming such impact is not already virtually unavoidable [Hansen et al. 1981] without limiting material standards to less than the current level of highly industrialized Western European societies. In reality, of course, there are non-energy constraints which make it unlikely that a world modelled on the FRG will ever become a reality; some students of development would even consider it undesirable. And even without those constraints, the extension of the Western urban/industrial culture to a global scale would be more likely to represent the imposition of ethnic and cultural biases on Third World peoples than a legitimate meeting of their aspirations for a dignified, autonomous, and self-reliant life. Thus if we are to obtain a more realistic profile of possible long-term world energy prospects, we need more sophisticated and realistic scenarios of world development, taking account of more modern philosophies (as espoused by such bodies as the Brandt Commission) and disaggregating the world enough to avoid the worst of the distortions introduced by the conservatisms (mentioned in the previous paragraph) inherent in our FRG energy budget.

We begin on a global scale by reviewing two scenarios for 2030: IIASA's "low" case[15] and Colombo & Bernadini's (1979) "low energy 2030" illustration. These, like their predecessors (e.g., World Energy Conference 1977; WAES 1977), are based on only the sketchiest analysis of the technical potential for cost-effective technical efficiency improvements, and their demand estimates must therefore be considered out of proportion. But they do provide a helpful background for improved estimates of the combined impact of population growth, alternative economic development strategies, and energy-efficient technologies on long-term global energy demand. It is therefore useful to

---

15. Because of the schedule of our own analytic work, the IIASA data used here are preliminary, and some differ modestly from the finally published version (IIASA 1981), but the differences are not important to our argument. We relied on Häfele (1979) and IIASA (1978), which gave a total primary demand for the "low" scenario of 25.7 TW rather than the final value of 22.4 (IIASA 1981). We also rely on the Colombo & Bernadini scenario in its original form (1979) rather than IIASA's (1981) discussion of it.

review them and contrast their different approaches to world economic development.

2.4.1. IIASA "LOW" SCENARIO

This projection—curiously named in view of its 3.35-fold increase in world primary energy use (excluding feedstocks) between 1975 and 2030 (7.66 to 25.7 TW)—assumes that:

- energy use is demand-led, not supply-constrained;
- the economic system and predominant lifestyle of the leading industrial countries remains essentially invariant over time save in intensity;
- this system will diffuse to all other countries over time; and
- the energy efficiency of most end-use technologies will not change substantially from 1975 values.

These assumptions have been given operational meaning by the following approach:

- GDP (and to some extent its rate of growth) is taken as the tacit measure of welfare;
- GDP growth rates are assumed for seven regions of the world based on population forecasts (Table 2.13) and on a qualitative extrapolation of past trends in world economic development before the oil shock (Table 2.14);
- for each region, historic variations of aggregate energy/GDP coefficient as a function of GDP or personal income per capita are used for projecting the future energy intensity of economic activity as a function of stage of development—tacitly assuming an identical development path for all countries in the future as for the most industrialized countries in the past; and
- substitution amongst primary fuels and the assumed increase in electrification of supply are also subsumed in these variations of energy/GDP ratios.

In principle, the demand forecasts were disaggregated and included some efficiency improvements. But as mentioned in Section 2.1, the model required many unavailable input data (2170 numbers to calculate demand for two dates and seven

Table 2.13 *IIASA population assumptions by region (millions)*

Region	1975	2000	2030
Developed regions	1161	1400	1562
I. North America	236	274	315
II. USSR and Eastern Europe	363	436	480
III. OECD Europe, Japan, Australia, NZ, SAfrica, Israel	562	680	767
Developing regions	2792	4682	6414
IV. Latin America	320	575	797
V. Africa except N & SAfr.; S & E Asia	1443	2548	3553
VI. Middle East and North Africa	132	247	353
VII. Centrally planned Asia	897	1312	1711
World total	3953	6082	7976

SOURCE: Keyfitz 1977.

Table 2.14 *Assumed growth of Gross Domestic Product in the IIASA (1978) "low" scenario*

	GDP per capita (1975 $)			Absolute GDP/1975 level (calculated from growth rates & Table 2.13)	
	1975	average growth rate (%/y)			
Region	level	1975–2000	2000–2030	2000	2030
I	7046	1.7	0.7	1.78	2.52
II	3416	3.1	1.9	2.61	5.08
III	4259	1.7	0.9	1.85	2.73
IV	1066	1.6	1.9	2.68	6.57
V	239	1.7	1.4	2.70	5.73
VI	1429	2.4	1.2	3.41	6.99
VII	352	1.6	1.4	2.18	4.33
world	1640	1.30	0.93	2.13	3.69
I+II+III	4546	2.04	1.19	2.01	3.21
all LDCs	426	1.86	1.53	2.67	5.78

ratio of GDP/capita for (I+II+III)/(IV+V+VI+VII):
 1975 = 10.7; 2030 = 10.1

regions) and was designed for France, not for a diverse world. These features forced the operators to make such crude simplifying assumptions that the apparent sophistication of the model could not be realized (Meadows 1981).

The use of energy/GDP ratios embodies two major methodological deficiencies beyond mere aggregation or the doubtful meaningfulness of GDP. Energy/GDP can be expressed as the product of *energy service intensity per unit of GDP* times *primary energy intensity per unit of delivered energy services*. The IIASA estimates of the latter term, as noted earlier, are far higher (less efficient) than (*a*) is econometrically consistent with the price of the assumed supply system or (*b*) would result from applying available, cost-effective efficiency improvements. This deficiency can be roughly quantified by uniformly applying to the IIASA energy intensities the efficiency improvements documented earlier for the FRG (Tables 2.9 and 2.10, "2030" case). This method ignores the effects of growth on composition of energy service demands, and therefore on total feasible efficiency improvements, but it is a useful approximation. Table 2.15 shows that on this basis, world energy use in 2030 would be about 13% *below* its 1975 level, rather than more than trebled as IIASA conjectured. Based on the final-draft IIASA data (1981), efficiency-corrected demand would be 5.9 TW, not 6.7 TW as shown.

Table 2.15 *Illustrative impact of aggregated efficiency improvements on IIASA (1978) "low"-scenario projections of primary energy demand (TW)*

Region	Primary energy use (excluding feedstocks)		
	1975	2030: IIASA "low"	2030: same + efficiency[a]
I	2.50	4.95	1.29
II	1.81	6.48	1.68
III	2.04	5.29	1.38
IV	0.32[b]	2.47	0.64
V	0.32[b]	2.84	0.74
VI	0.11[b]	1.24	0.32
VII	0.56[b]	2.40	0.62
world	7.66[b]	25.67	6.67
I+II+III	6.35	16.72	4.35
all LDCs	1.31[b]	8.95	2.32

a. Using the "2030" FRG improvement (×0.26) from Table 2.10. (Double-counting of IIASA's minor efficiency improvements is roughly equivalent to the further gain represented by our "present technical limit" case.)
b. The IIASA data for developing countries undercount noncommercial fuel use.

The second main methodological deficiency of the IIASA scenarios is implicit in the assumed pattern of variations of E/GDP ratio. This pattern, assumed to represent a "fated" trajectory for any nation following the development process, actually shows only the history of one extremely aggregated statistic for the presently industrialized countries at various stages of their development—which took place under unique circumstances giving rise to a far different optimization of factors of production than is the case today, let alone in the future. In that historic pattern, the early stages of industrialization, in which the urban/industrial infrastructure was built, showed a high relative demand for energy services per unit of GDP, followed by a lowering of this ratio as growth shifted more to information-intensive and service activities. The IIASA scenarios assume this pattern must also be repeated and continued for the next 50 years by the developing countries.

This deterministic assumption may well have been prompted by the IIASA analysts' sincere concern for bettering the living conditions of the world's poor. However, alternative approaches to development advocated by many Third World analysts (e.g., Reddy 1980) are increasingly finding official favor if only for the pragmatic reason that the "old development" has clearly not delivered what it promised. It now seems increasingly doubtful whether the IIASA development model could successfully provide the hoped-for humanitarian aid, because of limits on biological carrying capacity (President's Council on Environmental Quality & Department of State 1980, 1981; Lovins 1976), problems of land tenure, the social malaise that seems to accompany urbanization, the limited employment that can be generated in cities, and the dislocations caused by integrating local markets into the global economy.

2.4.2. THE COLOMBO/BERNADINI SCENARIO

In many respects, then, the well-meant IIASA scenarios embody a culturally biased approach to development, quite out of touch with today's global realities. Recognizing this impoverishment, Colombo and Bernadini (1979) set out to describe a more up-to-date development scenario, aimed basically at increasing the standard of living for the rural population (most of the world's poor) *where they now live* rather than dislocating them into destitute urban slums on an ever-increasing scale. To achieve this goal, regional and local economies with fewer dependencies on faraway, uncontrollable markets would need to be encouraged, and cultural diversity would need to be respected rather than homogenized. The availability of local, renewable energy sources (Chapter 4) that can power industry and support daily life virtually anywhere would of course greatly facilitate this approach.

The impact of such a decentralized, diverse development strategy on the demand for energy services would be considerable. Colombo and Bernadini identify the following qualitative effects:

- Much shorter travel distances within rural settlements would make walking, bicycling, and other low-energy modes convenient for most trips—a pattern that is now re-evolving in some Western countries.
- Greater living space and access to nature in and around settlements would obviate the need for much of the long-distance "escape" travel of urban dwellers and aid in maintaining psychic and biotic equilibrium.
- Greater self-sufficiency of local economies would reduce demand for long-distance freight transport.
- Rural self-reliance in food would reduce the need for elaborate storage, distribution, processing and packaging industries.
- The more ample supply of time and the lower transaction costs would permit, without hardship or drudgery, a partial substitution of labor for capital and mechanical energy.
- More closely knit rural societies would be better able to share equipment and appliances, leading to a decreased stock of embedded energy.
- Farming and forestry would need less energy- and capital-intensive mechanization and fewer energy-intensive chemicals, yet with modern bio-intensive techniques could still achieve higher yields, lower effluents, lower pest losses, and better integration with other productive activities.

- Industrial production would benefit from local structural materials, less transport infrastructure, reduced packaging requirements, and easier recycling. The energy demands of the steel, cement, paper, nonferrous metal, and petrochemical sectors could thus be substantially reduced.
- Energy demands in the chemical industry would also be reduced by greater substitution of natural products for synthetic fibers and structural materials.
- Greater product durability, easier repair (owing to more local production), lower mechanization, and in some cases a different product mix would somewhat reduce energy demands in the consumer goods and engineering sectors.

Focusing on these fundamental necessities common to all societies (food, clothing, shelter, health, culture), Colombo and Bernadini calculate (1979), by a case study of how energy is used in different settlement patterns, that a decentralized rural-based development pattern reduces the total need for energy services by about 2.8-fold compared with a highly urbanized pattern. They then construct a scenario for future energy service demand in which the majority of the expected increase in world population settles or remains in rural communities, and works in local economies. The assumed absolute size of the urban population increases, but the share of the world population that lives in urban settings stays constant at 30% or so, as shown in Table 2.16, saving energy otherwise needed for super-urbanization.

This less centralized growth pattern leads to an average 30% decrease in the energy/GDP ratio, and a world primary energy demand of 16 TW-y/y in 2030, without the technical improvements in energy efficiency described earlier.

2.4.3. An efficiency scenario

The foregoing discussion permits us to construct an illustrative scenario which incorporates both the development path of Colombo and Bernadini (1979) and the economically efficient energy use described in Section 2.3. Our starting points will be population and GDP growth, energy service demand per unit of GDP (structural energy service intensity), and technical energy efficiency of providing those services, as a function of time.

Table 2.16 *Possible urban populations in 2030*

Region	1975		2030	
	millions	% of total pop.	millions	% of total pop.
I	149	63	150	51
II	171	47	216	45
III	259	46	340	44
IV	131	41	279	35
V	274	19	782	22
VI	40	30	131	37
VII	215	24	479	28
world	1239	31	2377	29

SOURCE: Colombo & Bernadini 1979.

As time markers, we use:

- 1975, chosen for easy comparability with the other scenarios;
- 2000, marking the approximate end of the period in which developments are still significantly determined by near-term constraints, and permitting the incorporation of national studies for some important countries;
- 2030, approximately the time by which all major capital stocks (except buildings) should have turned over essentially completely; and
- 2080, by which population and economic activity may have reached a micro-variable, macro-stable state (Daly 1978) from which a perspective can be gained on the very-long-term fossil-fuel burn.

Until 2030, we use the IIASA/Colombo & Bernadini population projections, and thereafter we assume a world population stable at 8 billion. This asymptote is approximately achievable as the age structure stabilizes (Keyfitz 1977).

For greater ease of comparison, we assume the 1975–2030 GDP growth rates of the IIASA (1979) "low" scenario (Table 2.14)—not significantly different from those used by Colombo and Bernadini. GDP growth is understood in this context not as an indicator of welfare but as an aggregated surrogate for energy service demand. Since the ratio of average per capita GDP between the developed and developing countries is still 10:1 in 2030 under the IIASA growth rates, we assume for 2030–2080 a GDP growth rate of 1.1%/y for developing and zero for the developed countries, which will by then have average per capita GDP of approximately $10,900/y (1975 $), or 55% above the 1975 North American level.

We assume that in the industrialized countries, the structural energy service intensity per unit of GDP will be 0.8 of its 1975 value in the year 2000, 0.65 in the year 2030 (consistent with the Colombo & Bernadini pattern of settlements [Table 2.16] and of economic geography), and 0.5 in the year 2080, by which time purely material growth should have essentially saturated. These values are probably too high: a steeper decline can be expected for the FRG, for example, under purely urban/industrial development patterns, due to the international division of labor in the world economy, since energy-intensive, standardized secondary industries would not be able to compete against imports from young industrial or Third World countries such as Japan, Spain, or Brazil. Instead, the FRG would have to use its comparative advantage in R&D-intensive activities, which tend to have very low energy service intensities (IWW 1977). The combined effect of the intra-industrial and industrial-to-service shifts which now characterize the highly industrialized economies has been estimated (Krause et al. 1980) to lower the FRG's energy/GDP ratio by 40–46% by 2030. Our assumed relative energy service intensity for 2000 is an interpolation, and that for 2080 is an extrapolation consistent with Colombo's pattern.

For the developing countries, on the other hand, we assume that the relative energy service intensity will be 1.1 times the 1975 level in 2000, 0.9 in 2030, and 0.7 in 2080. This set of values is intended to reflect the opposing trends of less centralized development on the one hand and of continuous buildup of infrastructure and capital stock on the other. Again, the 2030 figure is based on Colombo and Bernadini's settlement pattern (Table 2.16)—and is roughly India's expectation for ca. 1982–95.

Our final set of assumptions concerns technical energy efficiency: the primary energy needed to provide a unit of energy service. Based on our earlier FRG discussion, and all the conservatisms it embodies, we take this coefficient, expressed again as an index relative to average 1975 levels, to be 0.5 in the year 2000, 0.26 in 2030, and 0.18 in 2080. The value for 2000 reflects the rate of introduction of efficiency improvements found feasible in the best available national studies, such as those by SERI (1981), Olivier et al. (1981), Leach et al. (1979), and Krause (1980). The value for 2030 corresponds to the technical potential of presently available and cost-effective efficiency improvements for the FRG "2030" case (Section 2.3 above), and is more pessimistic than Olivier's UK analysis (1981), whose assumed implementation rates have been exceeded in recent years. Insofar as other countries had a lower initial technical efficiency than the FRG, or faster population growth, or faster industrial construction and other acquisition of infrastructure, our wider use of the FRG results underestimates the potential for improvement. The 2080 value reflects our FRG analysis of the "present technical limit" (Table 2.10) for available, cost-effective measures fully applied.

It might be objected that coefficients derived from a static analysis of energy use in the FRG would not necessarily fit the structural composition of energy services and their thermodynamic requirements in other economies at other stages of development. This is perhaps inevitable in any calculation of this approximate and illustrative character. But the average technical savings in the industrial and transport sectors are similar in magnitude to each other (Table 2.9). Space conditioning, where the fractional saving by 2030 is about twice as large in the FRG, is not necessary in many populous parts of the world; and to the extent that electricity-specific functions (for which the long-term efficiency improvement is somewhat smaller than for other functions) increase their share of total energy service demands, the impact on our efficiency coefficients would be heavily buffered by the high degree of overelectrification in the 1973 FRG economy (see Chapter 3 below). Displacement of electricity from economically inappropriate functions, notably space and water heating, could equalize the actual potential for efficiency gains among all end-use categories. Electricity-specific applications also represent the smallest fraction of the end-use spectrum anyhow. For these reasons, the technical savings derived from our FRG analysis for 2030 can be considered an acceptable approximation for other economies too, and reflects multiple conservatisms.

Our calculation does not distinguish between the speed of deploying efficiency improvements in developed and in developing countries. While we consider this a conservatism in view of the latter countries' faster dilution or replacement of inefficient capital stocks, this assumption reflects also their dependence on many technology imports from countries that might be inclined to sell obsolete equipment to the Third World rather than selling plants that embody the best present art. Based on these assumptions, we calculate the result of our illustrative efficiency-and-"new-development" scenario to be as shown in Table 2.17 and Figure 2.6.

Most remarkably, despite a 4.6-fold increase in global economic activity during 1975-2080, and despite a doubling of world population, total energy needs for all three future years are *below* the 1975 level, dropping over the next century to less than half the present level. (The above figures are based on preliminary [1978] IIASA data which were revised downward 13% in final publication [1981], as noted above.) The 2030 estimate of 5.2 TW derived here is five times lower than the so-called IIASA "low" scenario (Häfele 1979). The additional use of "present technical limit" coefficients, even with another 24% increase in Gross World Product, yields total primary energy demand 6.3 times lower than the IIASA (1981) "low" scenario value for 2030.

Thus a more refined treatment than the simple *Gedankenexperiment* at the beginning of this section only underscores once more that estimates of future world energy needs published so far are scarcely beginning to take advantage of financially and socially attractive opportunities for meeting human needs with an elegant economy of means.

Table 2.17 Possible world primary energy demand (excluding feedstocks) with efficient use, strong economic growth, and constant urban fraction

Region	1975 demand (TW)	GDP/1975 GDP			Energy service intensity ÷ 1975 level			Technical energy efficiency ÷ 1975 level			Primary demand (TW)[a]		
		2000	2030	2080	2000	2030	2080	2000	2030	2080	2000	2030	2080
I	2.50	1.78	2.52	2.52	0.8	0.65	0.5	0.5	0.26	0.18	1.78	1.06	0.57
II	1.81	2.61	5.08	5.08	0.8	0.65	0.5	0.5	0.26	0.18	1.89	1.55	0.83
III	2.04	1.85	2.73	2.73	0.8	0.65	0.5	0.5	0.26	0.18	1.51	0.94	0.50
I+II+III	6.35	2.01	3.21	3.21	0.8	0.65	0.5	0.5	0.26	0.18	5.18	3.55	1.90
LDCs	1.31	2.67	5.78	9.98	1.1	0.9	0.7	0.5	0.26	0.18	1.89	1.68	1.65
world	7.66	2.13	3.69	4.57							7.07	5.23	3.55
index	1.00										0.92	0.68	0.46

a. May not exactly equal column products due to rounding.

Figure 2.6 *A global energy-efficient scenario (by regions) with strong economic growth and constant urban fraction*

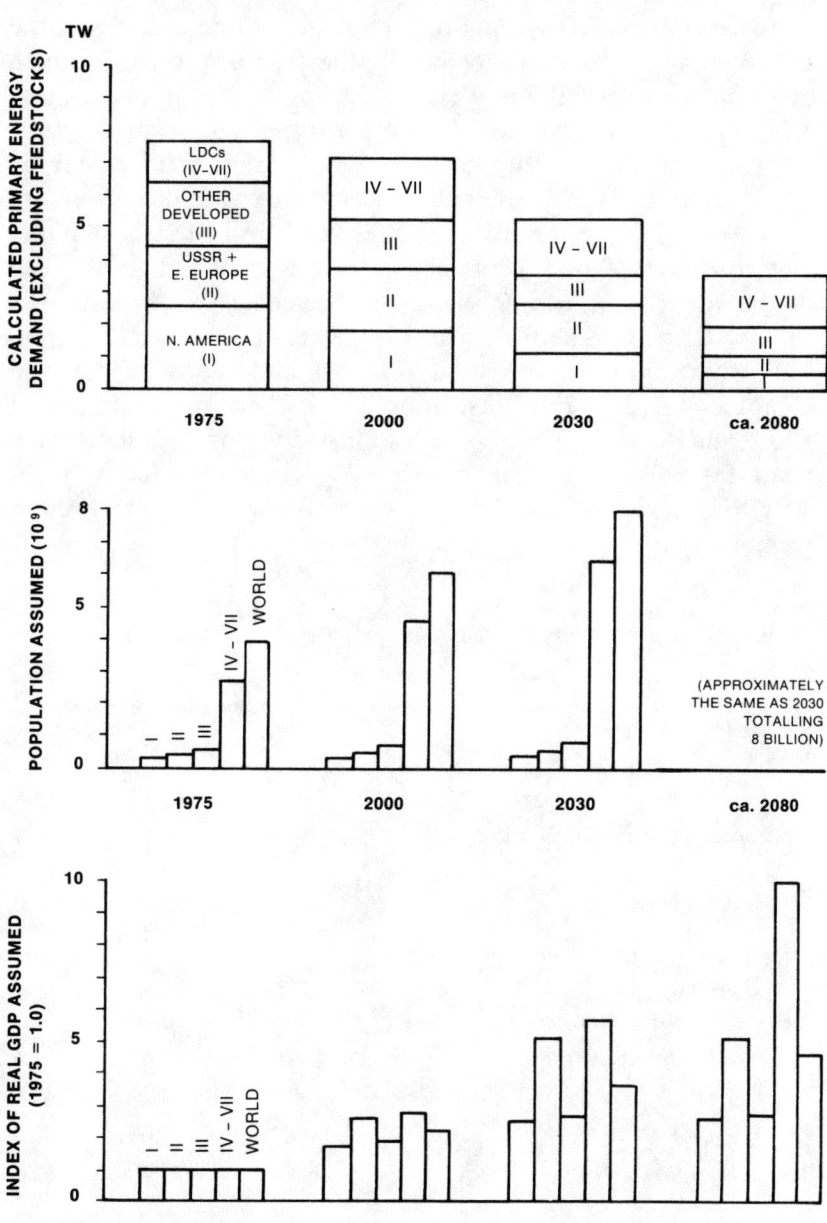

2.4.4. Implications for the European Economic Community

For specificity in drawing out the regional implications of the above scenario, we consider briefly the case of the EEC, which accounts for over 75% of Western Europe's primary energy use. EEC member states (with the temporary exception of the UK) are also dependent on imports for 55% of EEC primary energy needs, and nearly 90% of that imported energy is oil. There are no major fossil-fuel resources in Western Europe that could alleviate this dependence. There is, however, a vast "energy source" in the form of mineable energy inefficiencies which could be rapidly tapped. (There is also a large renewable energy potential, which we consider in a regional and global context in Chapters 4 and 7.) For illustration, we shall now reproduce in an EEC context the scenario exercise just done on a global scale, again tapping not the full "resource" of energy productivity but only the "reserves" that can be won economically with present technology.

Table 2.18 *Population assumptions for Western Europe within Region III*

	Population (millions)		
Area	1975	2000	2030
Western Europe	404.3	462.7	514.8
European Economic Community	258.0	295.6	314.4
France	52.9	63.4	69.6
FR Germany	61.7	57.6	50.0
Italy	55.0	63.8	67.4
United Kingdom	56.4	64.9	70.5
other	32.0	38.4	42.6
Rest of Northern Europe	31.3	33.6	35.1
Rest of Southern Europe	115.0	141.6	179.6
Japan	111.9	137.2	143.6
Rest of Region III	44.9	72.0	94.2
Region III total	561.1	671.9	752.7

The population assumptions in Table 2.18 correspond to Keyfitz's (1977) projections used also by IIASA, corrected in the case of the FRG to agree with national official projections (Enquête-Kommission 1980). The energy intensity of the EEC's Gross Regional Product differs widely among member states: it is almost twice as high, for example, in the UK as in France or the FRG, for historical reasons discussed by Colombo and Bernadini (1979). The future development of energy intensity per unit of economic activity is treated here in a manner analogous to that used in the global scenario. As for per capita GDP, we have used the values of Colombo and Bernadini (1979), which are in good agreement with those of the IIASA "low" scenario and with assumptions used in national energy studies (e.g., Enquête-Kommission 1980; Leach et al. 1979). The technical scope for efficiency improvements is again applied on a relative basis rather than as an absolute goal of kW/cap demand, since regional and climatic differences cannot otherwise be taken into account at all by this method. This approach is liable to underestimate the actual potential for technical savings, especially in the case of the UK (Olivier et al. 1981).

The results of the EEC exercise are summarized in Table 2.19. Despite the average 2.4-fold increase in economic activity, the total EEC primary energy use can be reduced by available, cost-effective technologies to 43% of its 1975 level. This result is all the more important since the 57% reduction in energy use just about equals the 55% import dependence of the EEC. Thus, even without any increased contributions from renewable sources, if present rates of domestic fuel extraction were merely maintained, the EEC would already gain energy independence—provided it did not first squander its money, technical skills, managerial attention, and political tolerance on relatively ineffective supply substitution technologies.

Having explored *how much* energy might (given economically efficient investments) be needed in the future for economic dynamism and global equity, we shall next consider *what kinds* of energy these needs represent, in order to be able thereafter to evaluate *what sources* can supply those amounts and forms of

Table 2.19 *Possible future primary energy demand (excluding feedstocks) in the EEC assuming rapid economic growth and economically efficient energy use*

Country	1975 demand (TW)	1975 GDP[a]	ec. growth rate/cap[b] 1975–2030	GDP (2030[c] ÷1975)	energy service intensity 2030÷1975	technical energy efficiency 2030÷1975	2030 demand (TW)
France	0.207	259.3	1.36	2.80	0.65	0.26	0.098
FR Germany	0.378	345.4	1.40	1.75	0.65	0.26	0.112
Italy	0.167	138.9	1.72	3.17	0.65	0.26	0.089
UK	0.279	178.6	1.14	2.35	0.65	0.26	0.111
Other	0.161	151.8	1.25	2.65	0.65	0.26	0.072
EEC total	1.13	1074.0	1.36	2.41			0.482
index	1.00						0.43

a. In trillion US$ (1975 $) at 1973 world market exchange rates.
b. In average percent per year compounded continuously.
c. Calculated using the population forecasts in Table 2.18.

energy at least cost. This will enable us to assess the potential contribution of renewable energy sources and hence to estimate *how much fossil fuel* will need to be burned for how long. That in turn will give some perspective on the prospects for reducing climatic risks through an economically efficient energy strategy on a national, regional, and global scale.

3
What Kinds of Energy Will We Need?

3.1. Matching heterogeneous end-use needs

Most analyses of the future energy system assume, in form or in practical effect or both, that the energy problem is where to get more energy—of any kind, from any source, at any price—to meet extrapolated *homogeneous* demands. But there are many different forms of energy whose different prices and qualities suit them to different uses. Further, there is no demand for energy *per se*—barrels of oil or raw kilowatt-hours—because energy is only an intermediate good, a means to an end. What people want is instead end-use energy *services,* the function which the final energy-converting device delivers to the users, such as comfort, light, mobility, TV reception, ability to smelt alumina or bake bread or run a sewing machine. It is therefore sensible to redefine the energy problem to be how to provide the *amount, type, and source of energy that will supply each desired end-use service at least cost.*[1]

Energy services are highly heterogeneous. If energy of superfluously high quality is used for a low-grade task, the mismatch will in general result in a low Second Law thermodynamic efficiency *and* a low economic efficiency. Least-cost solutions tend,

---

1. Here, as throughout this book, we define "cost" narrowly to mean "private internal cost," ignoring externalities and social costs which would only strengthen the comparison. This definition does not mean that we consider the externalities unimportant.

to first order, to be thermodynamically elegant solutions, too. For example, an extremely efficient heat pump powered by electricity from a newly ordered nuclear power station reflects a profound mismatch of source to task—degrading a reaction temperature (190 MeV/fission) of several trillion °C, via steam at hundreds of °C and infinite-temperature mechanical work and electricity, to provide 20°C heat in the room. It correspondingly reflects an economic mismatch: it uses a form of energy (marginal electricity) so expensive that even if used at maximum practical efficiency (the heat pump) it is grossly uneconomic compared to simpler alternatives, notably thermal efficiency improvements and passive (or even active) solar heat (Krause et al. 1980; Olivier et al. 1981; SERI 1981; Lovins & Lovins 1981; Lovins 1981d).

A logical starting point for designing energy supply systems is an analysis of what tasks the energy is ultimately to do. Such end-use structures can be indirectly derived (Lovins 1977, 1978) from available statistics on end-use and on the temperature spectrum of process heat requirements (InterTechnology 1977). Approximate results for representative industrial countries (ca. 1975) are shown in Table 3.1 and Figure 3.1. The data show the forms of energy *required* to run present processes at present rates, not the forms currently *supplied,* which are often of higher quality—as when fossil fuel is burned to provide steam at much lower temperature, or water is heated with electricity.

Comparison of the electricity-specific requirements with the electrical supplies (which have in general increased rapidly since 1975) illustrate a principle for which no exception has yet been found: that the electricity now supplied substantially exceeds that required for the premium uses that can justify the very high cost (>$20/GJ at the margin) of this special form of energy. The excess is even greater—typically severalfold—if present supplies are instead compared with what the electricity-specific demands *would* be after technical efficiency improvements which cost less than additional electricity (Lovins & Lovins 1980; Lovins 1981d). In contrast, most high-energy scenarios *assume*, without analytic foundation, greatly increased electrification, simply for a convenient match with assumed traditional sources—even though end-use demand both now and in the long term is overwhelmingly (today over 90%) for heating and for non-rail vehicles, both uses in which marginal electricity is grossly uneconomic (ibid.; Lovins 1978). Worse yet, this uneconomic increase in the electrical share of delivered energy causes an increase three times as large in primary energy demands, expressed mainly as additional fossil-fuel burn and hence as increments to the global CO_2 inventory. But from the

Table 3.1 *Percentage of total delivered energy (heat-supplied basis) needed in various forms in selected industrial countries around 1975*[a]

Form required	US	Japan	Sweden	UK	France	FRG	av. W. Europe
Total heat	58	68	71	66	61	75	71
<100°C	35	22	48	55	36	50	45
100–600°C	15	31	14	6	14	12	13
>600°C	8	15	9	5	11	13	13
Vehicular liqs.	34	20	19	26	29	18	22
Elec.-specific[b]	8	12	10	8	10	7	7
Indus. drive	5	7	6	4	6	4	4
Other electr.[b]	3	5	4	4	4	3	3
(Elec. supply)[c]	(13)	(16)	(18)	(14)	(12)	(13)	(11)

a. Energy now supplied for thermal tasks may be of higher quality than shown, as is implied by the last row (which shows the fraction of delivered energy supplied in the form of electricity). Temperature spectrum of process heat (InterTechnology 1977) assumes preheat from ambient with recuperated output heat, and is weighted for national industrial structure. Derivations and references in Lovins (1978) except for the UK (Olivier et al. 1981) and Japan, estimated from OECD statistics and from A. Doernberg, pp. 56–81 in Dunkerly (1978). Note that most of the medium-temperature heat is not very hot: even in Japan, 25/31 of the 100–600°C heat requirement is 100–315°C. The UN Economic Commission for Europe (Geneva) is currently collecting updated data on end-use structure as the basis for a major conference in Yugoslavia in autumn 1982; the ECE contact is Dr. Amasa Bishop.

b. Lights, electronics and telecommunications, electrochemistry, electrometallurgy, household non-thermal appliances, electric rail, arc-welding, etc. Thus all non-thermal uses of electricity are assumed here to be electricity-specific though some (notably lighting and drive) can be efficiently—sometimes more efficiently—done by other means (gas mantles and hydraulics respectively).

c. Fraction of delivered energy supplied in the form of electricity. The excess over electricity-specific requirements is electricity used for heating and cooling. Additional electricity could only be so used.

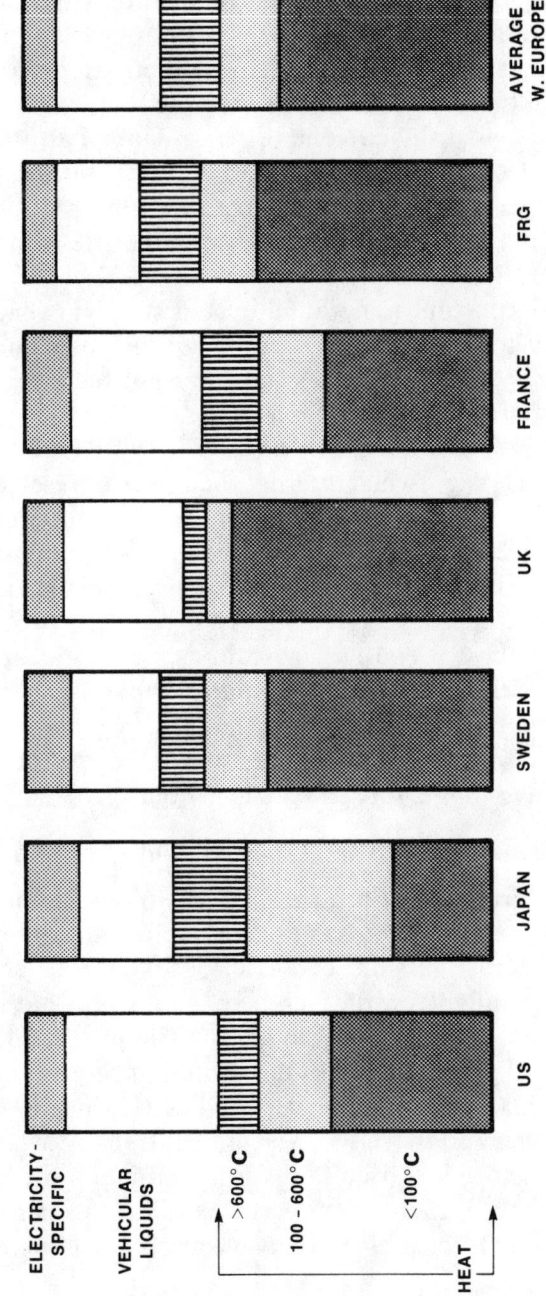

Figure 3.1 *End-use structure of selected industrial countries, ca. 1975*

economically conservative perspective of supplying each end-use need at lowest private internal cost (or, in plain language, doing each task in the cheapest way), new thermal power stations, regardless of their fuel, are so uneconomic that it is cheaper for a country that has just built one to write it off than to operate it (Lovins & Lovins 1980:48). This suggests serious economic inconsistencies in scenarios involving new power stations of any kind, particularly central steam-raising plants (Lovins 1981b, c, d; Sant 1981; SERI 1981).

It is also important to note that just as supplying energy in the right *quality* for each task can minimize the costs and losses of conversion, so supplying energy at the right *scale* for each task can minimize the costs and losses of distribution. This important issue is analyzed elsewhere (Lovins 1977, 1978, 1978b; Lovins & Lovins 1981). A growing international literature strongly suggests that gross economies of scale in supply are often more than counterbalanced by previously uncounted diseconomies of scale: what matters is the net, not gross, economies, and the savings expected from centralizing energy systems often prove illusory. Matching supply to end-use in scale as well as in quality provides much of the economic advantage of the supply strategy proposed in Chapter 4. It also means that energy (chiefly fossil fuel) which would otherwise have been lost in distribution can be saved, thus avoiding a further CO_2 contribution.

3.2. Evolving thermodynamic structure of end-use needs

For long-term supply projections, it is of course not only the historic but also the future end-use structure that is important. Although accurate versions of Table 3.1 for several decades from now are not available, estimates exact enough for our needs here can be readily derived from the data in Chapter 2. In the FRG, for example, Table 2.9 implies the evolution shown in Table 3.2. Olivier et al. (1981), like earlier studies (Lovins 1978), find a similar pattern for the UK—a significant shift towards higher proportions needed in premium forms, although the *amount* of energy required in each form decreases. Thus forms which could entail higher CO_2 releases, notably electricity, should not become dominant.

Table 3.2 *Possible evolution of FRG end-use structure*

	Approximate % of end-use demand (excluding feedstocks) in:		
End-use category[a]	1973	2030	ca. 2080
Total heat	74	65	55
(of which <100°C)	(50)	(34)[c]	(26)[c]
Vehicular liquids	19[b]	21	28
Electricity-specific	7[a]	14[c]	17[c]

a. Correction the 1973 entry to show, as in Table 3.1, requirements for electricity rather than the fraction of delivered energy supplied in that form. (Note that Fig. 2.1 shows the latter). Heat figures include steel-making reductant.
b. Does not agree with Table 3.1 owing to rounding errors.
c. Corrected for electrical process heat (moved to the heat row), assuming that of the electricity used for process, ⅔ in the chemical sector and ¼ in "other industry" is for heat, and that a constant ¼ of post-preheat process heat in all industries (as in the US now [InterTechnology 1977]) is <100°C.

4
What Energy Sources Can We Sustainably Use?

4.1. Monolithic versus diverse/competitive strategies

The only energy sources free from significant climatic impact (given good management) are dispersed renewable sources (Holdren et al. 1980). Today the energy system is mainly non-renewable. It is outwardly diverse—different sources of oil, gas, and coal, different ways of burning them, different kinds of nuclear reactors and hydroelectric plants—but in outline it relies on a few monolithic energy systems. These offer such technical and logistical challenges that they have evolved major industries, technical institutions, financial arrangements, and political constituencies. This structure is in turn reflected in the structure and data of most formal models of the energy system: most accounting models, including those used by IIASA, simply cannot handle the bewildering array of non-fossil, non-nuclear technologies, most of which are therefore "excluded by assumption" (Meadows 1981). Such models can therefore be used to suggest that those sources cannot play a significant role, so they may attract still less official interest.

There is no single "renewable source" or "solar technology." The types and combination already known represent an immense diversity of type, integration opportunities, scale, suitable climate, appropriate application, technical sophistication, and accessibility to various potential users. This spectrum is itself the greatest strength of renewable energy sources. It means there is virtually bound to be one suited to any conceivable set of circumstances, rather than forcing users to conform to the imperatives of more rigid technologies.

But this very adaptability poses a formidable problem for the analyst, for it is not enough just to list generic technologies: most renewable sources work best, both technically and socially, when *matched* to a particular task, place, and user. Their cost and performance depend sensitively on their use and context, and cannot be accurately judged in isolation. Most analysts also find it unfamiliar, and therefore difficult, to consider the possibility of meeting the energy needs of a major industrial nation with a large number of diverse, often dispersed and relatively small, sources. Even though this is the way in which the national treasury is fed by many individually modest tax contributions, energy analysts are not used to thinking in this way. They like to deal in GW or TW, not W or kW. It is therefore common—though an egregious error—to dismiss as too small the contribution of any particular one of a myriad renewable sources, or to fault it for its unattractiveness if it is used to try to imitate the style or scale of conventional centralized systems.

This conceptual problem is real and thorny. It is not just an artifact of scientific sociology; nor is it merely the natural desire of people who have devoted their careers to technologies now in market difficulties to seek to shield their accomplishments from potentially rigorous competition; nor is it only analysts' craving for the orderly simplicity of single, nominal, neatly describable technologies rather than the near-chaos of an amorphous, unruly crowd. Although the inadequate treatment of renewable sources in much of the professional literature is doubtless related to these factors, it stems also from the sheer intellectual difficulty of coping with a flood of widely differing technologies, many changing with dizzying speed. The best designs have probably not been thought of yet, and fundamentally new and attractive concepts are still emerging frequently. Keeping up with progress in even one broad category of technical advance—say, lignocellulosic conversion, or direct photolysis, or solar ponds—can keep a team of able analysts fully occupied.

All these difficulties may incommode scholars more than the general public. A buyer in the marketplace may see ten or a hundred types of solar water heater available, and can choose one to put on the roof based on the best information at hand, without being put to the harder task of comparing *all* solar technologies under all circumstances both with each other and with non-solar options. A buyer can see the nearest trees without having to map the forest. We cannot proceed quite so simply, but will try without going to unmanageable length to set out concisely, with basic documentation, the reasons for the remarkable recent discovery, in the marketplace and in the energy policy community (Stobaugh & Yergin 1979), that the pessimistic early judgments of renewable resources did not begin to do justice to their potential.

The notion that renewable energy sources are incapable of playing an important role in long-term energy supply is nowadays confined to the uninformed. Somewhat more knowledgeable analysts, however, long held that although renewable sources were a major long-term option, they still required extensive and speculative development before they were available in significant numbers and diversity at attractive prices. This still common conclusion, too, has lately fallen victim to technological progress. Careful study of the best renewable sources already in or entering commercial service, matched in scale and energy quality to their tasks, and cost-effective at least at the margin, has shown that they are sufficient to meet virtually all the long-term energy needs in every industrial country so far studied, including among others the US (Sørensen 1980), Canada (Lovins 1976a; Brooks 1981), France (Groupe de Bellevue 1978), Sweden (Johansson & Steen 1980), Denmark (Meyer et al. 1977–79), and Japan (Tsuchiya 1980, 1981). (We consider below the least favorable cases—the UK and the FRG.)

This list of countries for which renewable energy supply scenarios have already been examined in sufficient detail to make a persuasive case is highly suggestive: it includes very densely populated and heavily industrialized countries with cold and cloudy climates, low biomass production, and little hydroelectricity. Nonetheless, while technical opportunities vary between and even within these countries, each has enough. Countries that are poor in fuels, such as Japan, can be singularly rich in renewable energy of many kinds (Tsuchiya 1980, 1981). This suggests similar inferences for other countries not yet studied in detail but clearly possessing a broadly similar range of opportunities owing to their varying combinations of insolation, biomass base, wind regime, hydro potential, and other energy flux resources. Further, although it is analytically convenient to study a country in isolation, there is much recent evidence that total renewable potential is often cheaper and more convenient if considered either at a regional scale (e.g., for Scandinavia [Sørensen 1981] or Western Europe [Caputo 1981]) or at a relatively local scale (SERI 1980; Craig et al. 1978; Reddy 1980; Pomerance et al. 1979; Glidden & High 1980). We shall return in Section 4.2.3 to the national, regional, and global adequacy of renewable sources, after further examining their nature and their technical, commercial, and economic status.

4.2. Renewable sources with inherently low climatic risk

The main classes of renewable sources considered here are: passive (preferably) or active solar heating and cooling (distinguished by the active systems' use of special collectors and mechanical circulation of a working fluid); specialized active collectors which achieve high temperatures; solar ponds; conversion of farm and forestry wastes (and possibly, in some special cases, of "energy crops" on otherwise unproductive land) to fluid fuels—mainly alcohols and pyrolysis oils—via thermochemical processes, fermentation and anaerobic digestion, acid and enzymatic hydrolysis, advanced alcohol-water separation techniques, and other processes; present large-scale and readily available small-scale hydroelectricity; wind turbines; photovoltaic cells, which convert sunlight directly to electricity; and occasionally some small-scale burning or gasification of wood and municipal wastes. Although this seems a simple enough schema, it actually embraces an enormous and rapidly evolving range of technologies (Sørensen 1979a; Lovins 1978b), some new and some merely reinvented (Butti & Perlin 1980). For example, wind turbines can not only generate electricity, but also provide mechanical work, hydraulic compression, or heat.

We omit here photochemical options other than bioconversion; tidal power, which is not very important; wavepower, which may become attractive in special circumstances but does not currently seem necessary; and geothermal energy, which is generally not renewable, although it may be locally quite useful. We also omit those proposed solar technologies which we consider unnecessary and neither economically nor ecologically attractive: "power towers" (large-scale solar-thermal-electric conversion—although the heliostat technology itself may be useful for some specialized high-temperature direct-heat applications); ocean-thermal-electric conversion; mariculture; monocultural biomass plantations; and solar power satellites (which work better if laid on the ground in Hamburg). We consider rather a diverse subset of all solar technologies called "soft technologies" (Lovins 1977, 1978b) and characterized by a match in scale and quality to their task and by relative understandability to the user (though they can, like a pocket calculator, be technically very sophisticated).

4.2.1. Technical and Commercial Status

The first obstacle to assessing which soft technologies are available and how they perform is that the best present art is highly dispersed among thousands of workers in dozens of countries. Many of these researchers have more ingenuity than technical credentials, and most are outside official R&D programs, which are often the last to know. Information flows to the conventional energy hardware community are slow and, in general, seriously deficient, especially since most official programs are attuned to central-electric systems—the recipient of about three-quarters of most major countries' energy R&D investments for the past few decades—whereas soft technologies provide all the forms of energy that end-users require, mainly heat and portable liquids. (The public is often further ahead in suspecting the true state of affairs, but lacks particulars.) Only people with extensive informal networks and the freedom to travel extensively to examine grassroots technical developments can expect to be aware of even a small fraction of what has been done. Despite our best efforts, we probably hear of only a few percent of the best work within a year of its occurrence.

To make matters worse, progress is too fast for even specialized newsletters to keep up with. Half the cost data in a review article (Lovins 1978b) had to be revised within four months of the first draft. Most performance and cost goals are bettered monthly. While the direction of change is generally favorable, its pace is so fast that conventional channels of publication are years behind, and even such short-lead-time efforts as *Soft Energy Notes* are on occasion up to a year behind communications on the telephone-and-typescript grapevine. The flow of exciting developments greatly outruns any staff's time to read it or room to publish it. The speed of progress can also become a real barrier to commercialization, as in photovoltaics, whose current "breakthrough-per-month" syndrome discourages investment in any particular manufacturing process, since it is bound to become quickly obsolete.

Soft technologies are commercially available in most industrial countries in all the categories listed above (Lovins 1978b; SERI 1981).[1] Some are being sold in large quantities: the two million solar collectors in Japan increased by $0.5 billion worth in 1980, and 1979–81 commercial commitments to windpower in the US probably total several billion dollars. Performance depends on the cleverness of the designer, the skill of the builder and installer, the technique of the user, and the suitability of the device to unique local circumstances. We therefore rely on data showing performance that is being routinely achieved with intelligent designs and can reasonably be expected to be replicable in widespread commercial practice. Such estimates, often with extra safety margins of conservatism built in, are exhaustively surveyed by SERI (1981) and Sørensen (1979a).

We restrict our analysis to technologies already in or entering commercial service, and exclude improvements that are not yet being tooled up for production. Several such developments, however, notably in photovoltaics and in the efficient conversion of cellulosic materials to alcohols, promise dramatic cost reduc-

---

1. There are minor exceptions, not important for our argument, in some lignocellulosic conversion and high-temperature process heat technologies which are demonstrated but still in advanced pilot development (see pp. 112–113, 123).

tions based on successful laboratory and pilot experiments. The US Department of Energy, for example, now expects photovoltaics without a cogeneration heat credit to compete on US utility grids by 1986, based only on conventional silicon technologies already in hand (Adler & Maycock 1981; Russell 1981; Maycock & Stirewalt 1981). The pace of experimental developments in the US, Europe, and Japan makes it seem highly likely that amorphous cells will do this even sooner.

Direct solar energy is intermittent, and concern is therefore often expressed about the need for energy storage. In fact, this is less of a problem in an energy-efficient renewable supply system than in a nonrenewable one. It is straightforward to store sensible or latent heat, and, as noted below, the storage volumes required are modest. Liquid fuels store themselves. Highly efficient energy use will ensure that, at least in the main industrial regions today (North American, Europe, Japan), and probably in others in the longer term too, the supply of electricity to the grid will be dominated by hydroelectric capacity, both large and small, permitting storage in the form of water behind dams, mainly existing ones. Indeed, because hydro, windpower, and photovoltaics work best at different times, integrating them actually yields a *more* reliable grid supply than conventional thermal plants (Kahn 1979; Sørensen 1979a). In contrast, conventional supply projections envisage a highly electrified economy, incurring serious and so far unresolved problems of how to store electricity in bulk—at best an extremely awkward and expensive task.

Another common misconception concerns the influence of clouds (Sørensen 1979a). Passive solar systems work particularly well in cloudy weather, photovoltaics and photosynthetic systems merely reduce their output in proportion to the reduced intensity of total insolation, and even non-focusing active solar heating systems can work well if designed for these conditions. This is because the diffusely scattered light, although somewhat reduced in intensity, is still energy of extremely high quality, with an effective color temperature of at least 3000°C. Much, even most, of this thermodynamic potential can be captured through the use of selective surfaces, which absorb well at visible wavelengths but radiate poorly in the infrared, and therefore, if thermally isolated, attain very high equilibrium temperatures. A selectivity (visible absorptivity ÷ infrared emissivity) >50 can be achieved by sputtering thin films, using well-established high-vacuum techniques similar to those used for coating optical lenses but far less critical. It can be easily calculated that a surface with a selectivity of 50, in a hard vacuum, will produce under load heat at 500-600°C on a cloudy winter day in Hamburg; and if, under those conditions, the liquid-metal coolant should stop flowing without a protective device to shield the absorber, the metal absorber plate would very probably melt. Even a modest selectivity of about eight—readily achieved with commercially available coatings, paints, electrodeposited surface treatments, and foils (costing \$3/m^2 [Raetz 1979])—permits even a single-glazed solar water heater to perform well in a northern German climate with total insolation of only 200–300 W/m^2 (20–30% of the summer sunny-day level), heating the water by 30°C at 45% and 57% First Law efficiencies respectively (Raetz 1979a). Likewise, through passive solar design, net space heating loads less than 3 kJ/DD-m^2 have been obtained (SERI 1981) in cold, cloudy climates—implying a heating load equivalent, for a 150-m^2 dwelling, to <140 kg of wood/y burned in a good stove, or to heating the house with one or two 40-watt dogs. By suitable design, therefore, the handicap of cloudiness can be overcome at reasonable cost.

Still another common misconception about soft technologies is that they require large land areas. Aerial photogrammetric surveys of cities in Sweden and the US have found ample unshaded, Equator-facing roof and wall area to collect (even by active devices alone) enough energy to heat the buildings and hot water, and usually with photovoltaics to cover all the electrical needs too, given modest efficiency gains. (Urban density generally *improves* solar economics.) Even completely renewable supply of all energy needs in land-poor Japan would probably need little or no extra land (Lovins 1980a).

Nor would renewable liquid fuels require diversion of farmland from food production. In fact, if vehicles are cost-effectively efficient, just the farm and forestry residues already produced in many countries as an otherwise unwanted by-product could, if efficiently converted to alcohols and pyrolysates, operate the transport sector. In the US, which currently uses a third of the world's energy, the technical efficiencies documented in Section 2.3 would reduce the US liquid fuel needs for transport from about 20 to about 5–6 EJ/y, and many analysts think that this much liquid fuel could be sustainably produced (with careful management) without increasing or displacing food and timber production (SERI 1981; OTA 1980). Even urban sources abound: just the pure, separated tree material sent to Los Angeles County landfills totals 3600–7200 T/d, with a heat content of the order of 1 GWt. But biomass wastes are also surprisingly large in other countries. The straw burned in French fields is equivalent to a tenth of total French primary energy consumption (Lewis 1980); the corresponding figure in Denmark is probably even larger; biomass residues for Europe as a whole probably exceed efficient transport needs (Caputo 1980, 1981). Even Japan is two-thirds forested and produces extensive timber and agricultural wastes, totalling (with municipal and organic industrial wastes) some 66 dry MT/y—enough, with modest efficiency improvements, to meet all national needs for transport and cooking with a bit left over for industrial process heat (Tsuchiya 1980, 1981).

Although it is possible to find processes which convert sugars, starches, or cellulose to liquid fuels at low process efficiencies or with large energy input requirements, and some analysts apparently assume that a small, zero, or negative net energy yield is unavoidable, careful attention to appropriate scale and effective use of by-products, provision of process heat by non-petroleum sources, and modern alcohol-water separation processes (highly efficient stills, freezing, chemical extractants, hydrophobic plastics, cellulosic adsorbants, etc.) can consistently yield large net energy gains (Ferchack & Pye 1981a; Patterson 1980). For example, the energy to distill ethanol to 190°PR has been reduced from the standard 55–100,000 BTU/US gal to 8–10,000 by advanced distillation or critical-fluid processes, and even anhydrous ethanol (which is not necessary for unblended fuel use) can now be produced within the same energy budget (Ferchack & Pye 1981a). Good process efficiencies today (weight yield from feedstock to alcohol) are typically 0.46–0.48 for glucose fermentation to ethanol (90-95% of the theoretical limit), 0.5–0.6 for acid hydrolysis of cellulose to glucose (Grethlein et al. 1980) or 0.99 with a solvent/enzymatic method (Ladisch et al. 1978), and hence about 0.4+ (Ferchack & Pye 1981; SERI 1981) for holocellulose to ethanol if the better processes now in pilot testing or becoming commercial are combined. Wang (1981) sees "no technical barriers" to scaling up an 85%-of-theoretical-yield laboratory process for direct cellulose-to-ethanol conversion. Thermochemical processes are also quite efficient: liquid-optimized pyrolysis of wood wastes to heavy oil, for example, has typical yields >0.5 (Lindström 1979), and the oil-plus-char yield (which can be slurried together) is typically 0.6–0.8, even in small units (Tatom et al. 1976). Wood gasification and subsequent catalytic shift to methanol yield approximately 0.40–0.48 (OTA 1980). Thus it is certainly not true, as Schmitz and Voss (1980) state, that "a 40% conversion efficiency of biomass to liquid fuels is not supported in the scientific literature." Alcohol blends (especially with butanol) and engine optimization can also burn cleaner and yield considerably better engine performance than with gasoline, thus largely or wholly compensating for the alcohols' lower energy density per liter.

As the availability of biomass feedstocks illustrates (SERI 1981; Jackson 1980; Jackson & Bender 1980, 1980a; Lovins & Lovins 1981a), the supply curve reflecting the cost (or the difficulty or nastiness) of soft technologies rises steeply and often discontinuously if they are used in the very large quantities to which inefficient energy use would drive them. They are relatively benign and attractive only if used on the shallowly sloping lower portion of the supply curve—the economically efficient region for supply-demand balance. For example, 8 TW of global soft-energy supply with a typical long-term end-use structure (Table 3.2) is straightforward. The 1–1.5 TW of electricity needed[2] can

---

2. This figure is probably too high. Olivier et al. (1981) found that 2.9× 1976 UK GDP required about 174 W/cap of electricity, implying 1.4 TW for 8 billion people but only about 0.3 TW at our assumed 2030 level of global GDP/cap.

be readily supplied by macro- and microhydro alone, but in practice would be extensively supplemented with wind and cogeneration and perhaps with photovoltaics. The 2–2.5 TW of liquids needed, derived from about 3–5 TW of farm and forestry wastes, represents a small fraction of readily collectable biomass residues from present harvesting.[3] The remaining heat requires little or no extra land area for direct solar collection—certainly far less than with conventional systems. A 16-TW soft supply scenario is still feasible but may be less than half as attractive, and may start to incur marginal land requirements—though still less than for hard technologies (Grenon 1975). But still higher demand levels, such as those of the IIASA scenarios, would increasingly incur a spurious "need" for "hard" solar technologies—which have never proved economic or necessary in any case in which efficiency was optimized first.

The same interaction between demand and supply occurs on the scale of designing individual renewable supply devices. This crucial point, which is ignored in virtually all official analyses, is illustrated by the Saskatchewan Conservation House mentioned in Section 2.3.1. Its salient characteristics are shown in Table 4.1, as measured in operations since December 1977 (Besant et al. 1978; Dumont et al. 1978). Why is such a small solar system big enough to cover the entire load, while most studies say a system 5–10 times as big would be needed to achieve partial coverage?

- The space-heating load is tiny, averaging one-third as big as the water-heating baseload. Thus less collector area and storage volume are needed.
- Although the house has light frame construction with low thermal mass (25 MJ/C°), it has a 100-hour time constant, so the peak loads for space heating are not only small, but smoothed out over a long time.
- Since the inside of the house is nearly in convective and radiative equilibrium, any point source of heat (e.g., one short

3. For comparison, the average annual global net primary productivity on land is about 76 TW (ca. 0.2% of insolation) and the land standing biomass is about 450 TW-y. Many residues can be collected cheaply (Lovins & Lovins 1981a).

Table 4.1 *Characteristics of the Saskatchewan Conservation House*

Site: Regina, Saskatchewan, 50.5°N lat.; average insolation 162 W/m^2; design temperature −34°C; 6000 Celsius degree-days/y against 21°C

Fabric: 2-story frame-construction box, 187 m^2 floorspace

    windows: 13.8 m^2 (86% west-southwest), double-glazed downstairs, triple-glazed upstairs, night shutters

    shell: heat-loss rate (W/m^2-C°ΔT) = 0.09 roof, 0.14 walls, 2.1–3.3 windows (open shutters), 0.33–0.43 windows (closed shutters)

    ventilation: uncontrolled infiltration <0.05 air change/h (ac/h) due to tight 0.15-mm vapor barrier; ample exchange via air-to-air heat exchanger with average efficiency 0.8, peak 0.93, materials cost ca. $100 (or ca. $30 without blower), average exchange rate 0.6 ac/h, maximum 0.8 ac/h.

    domestic hot water: load reduced by a third by a graywater recuperator

Measured performance: gross shell loss   43.5 GJ/y = 68 (99) W/C°ΔT with shutters open (closed)

    − net passive gain   14.2 GJ/y
    − people gain   4.6 GJ/y
    − appliance gain   19.6 GJ/y
    = net space heating load[a]   5.1 GJ/y = 162 W average = 1400 kWh/y

peak heating load: 5.45 (3.73) kW with shutters open (closed) @ 99C°ΔT

Active (evacuated-tube) solar collector system is big enough to cover ca. 100% of space & water heating load, without backup, using only 17.8 m^2 collectors (9.5% of floor area) and 11.7 m^3 water storage (2.8% of house volume)

a. Rosenfeld et al. (1980) give higher values: see note 7 in Chapter 2.

piece of uninsulated pipe) will heat the whole house evenly, so no heat distribution system is needed.

Thus the solar system can be much smaller and simpler than usual. Moreover, since solar water heating can be cost-effective by itself, that role can bear the whole fixed cost of the solar installation. The slight oversizing of the water-heating system to cover the residual space-heating load too incurs only a *marginal variable* cost for space-heating, and that cost is less (with good design and efficient construction) than the capital cost of a gas furnace or wood stove. Thus if the cost-effective efficiency improvements are done *first*, 100% active-solar heating is cheaper than partly solar heating, because the storage has become so small that it costs less than backup heating. (The 100% solar heating is then also cheaper than any other source of heat except passive solar heat or solar district heating.)

This example illustrates the powerful synergism between efficiency and cost-effective renewable energy systems. Other such examples abound: for example, if thermal efficiency measures reduce the space-heating load of a building to a low level, the residual load can often be completely covered (even in a severe climate) with a rather small add-on Equator-facing greenhouse containing adequate thermal mass. One can then put simple, even unglazed, flat-plate collector panels for domestic hot water in the upper part of the greenhouse *where they can never freeze,* thus avoiding the cost and complexity of antifreeze loops, heat exchangers, drain plugs, and other frost precautions.

Unit scale can also critically influence the size and performance of renewable energy systems by changing their basic physics, just as the small peak loads and long time constants of the Saskatchewan Conservation House changed the physics of its solar system. Solar district heating (Margen 1980) offers an example, relevant to solar retrofitting of urban or historic areas. A large water tank for heat storage, shared between tens or hundreds of dwellings, provides a large volume-to-surface ratio (hence low heat losses), low marginal cost per m^3, and a favorable ratio of variable to fixed costs. Its low cost permits the installation of several m^3 of storage volume per m^2 of collector area (compared to typically 0.5–0.8 in single houses with quasi-seasonal storage [Lovins 1978b]). That provides true seasonal storage, which in turn improves annual collector efficiency by providing a full summer load; encourages stratification or other temperature segregation of zones of the tank, further boosting efficiency and reducing heat losses; and permits one to face collectors east or west with little performance penalty. The net result is to reduce the total cost per unit of delivered heat by a factor of up to two—to levels competitive with C$8/bbl (US $1.3/GJ) wellhead oil in 1977 (Hollands & Orgill 1977), and certainly competitive with electric heat (OTA 1977).

4.2.2. ECONOMIC STATUS AT THE MARGIN

The preceding section lists only a few of the subtleties of assessing the costs of soft energy systems. Another is market structure: equivalent active solar systems, for example, can now be bought in the US for an installed system price of $250–500/m^2 collector area if the system is "packaged"—manufactured remotely, with the price marked up three or four times along the way—or, alternatively, for only $100–150/m^2 if the collector is "site-assembled"—that is, built on the house itself, with the price marked up only once. The difference in price, at least in the Northeastern US, where on-site assembly is common, is often at least a factor of two, just due to the difference of market structure (Worcester Polytechnic 1978). Indeed, there is ordinarily a scatter of at least twofold in the installed price of identical gas furnaces put in at the same time in apartments in the same part of California, depending on the skill and avarice of the contractor; so it is not surprising that solar hardware, which is more diverse and has a less mature marketing structure, should show an even wider price variation, requiring careful shopping for best buys.

Perhaps the most difficult determinant of soft technology cost to define is its complexity or simplicity of design. Overengineered, highly instrumented wind machines can easily cost ten times as much per kW as elegantly simple designs that are probably a good deal more durable and reliable. Greenhouses can be glazed either with tempered glass or (just as serviceably and far more cheaply) with special plastic and composite materials. Materials can be new or recycled, bought retail or wholesale (e.g., by a solar cooperative). The mass per unit area of solar collectors can vary by severalfold—or, with some rapidly emerging designs using durable plastic and metal film materials, by nearly a hundredfold—without greatly affecting durability or performance. Convenience is somewhat more measurable than simplicity, and our analysis assumes devices that are just as convenient and "hands-off" (and just as reliable) as conventional supply systems; but if we wished to assume even a slight degree of routine maintenance or tinkering, we could cut many soft-technology costs considerably.

Subject, however, to these caveats, we have relied on extensive analyses of comparative cost in a range of countries and conditions (SERI 1981; Lovins 1981b; Lovins & Lovins 1981; Olivier et al. 1981). They support in detail the finding that if properly designed soft technologies are fairly and symmetrically compared with their conventional competitors, the soft technologies consistently have lower capital cost, and several times lower delivered energy service price to the final user, than do synfuel or power plants which would otherwise have to be built to do the same tasks. Some soft technologies cost more and some cost less (mostly less) than today's oil; nearly all of them cost more than the efficiency improvements without which neither they nor any other supply systems makes sense; but they are all cheaper than their competitors. Though not cheap, soft technologies are cheaper than not having them.

A few years ago, this result was controversial, because soft technologies had until then been compared in cost only with dwindling, historically cheap, and often heavily subsidized fossil fuels. But those fuels' proposed replacements—synfuel and power plants, and in the shorter term such exotic delivery systems as LNG tankers and Arctic gas pipelines—were being compared in cost not with those relatively cheap old fuels, but only with each other, and they would have failed by a far wider margin the test applied to the soft technologies. More recently, as more numerous and disinterested economists have considered this question, the finding that the soft technologies are the second best buy—worse than efficiency, but better than the rest—has become rather widely accepted (Stobaugh & Yergin 1979; Sant 1979, 1981; SERI 1981; Southern California Edison Co. 1980; Swedish Energy R&D Commission 1980).

The level of detail needed to review these economic comparisons satisfactorily would fill a large book, and in a report of this length we can therefore only refer the reader to the original surveys (especially the authoritative treatment by SERI [1981]). But we emphasize the degree of conservatism which these comparisons embody. They do not assume a continuation of the consistent trend of the 1970s—that the real price of nonrenewable sources rose while that of renewables fell (or, at worst, rose far more slowly). They generally assume unrealistically low prices for nonrenewable systems; they use the same fixed charge rate for both (this discriminates against renewables, and ignores the cash flow advantages of shorter lead time and faster payback, both of which reduce working capital requirements per unit of capital intensity); and they ignore cheap solar designs (such as plastic-film collectors, collectors integrated into the fabric of the building, solar district heating, and—in our work [Lovins 1978b]—passive systems). Our comparison of, for example, active solar heating vs. nuclear-electric heat pumps (Lovins 1977, 1978c, 1979) also assumes an unrealistically efficient heat pump (coldest-day coefficient of performance of 2.5), and assumes that the baseload electricity price represents the average or peak price. By these and other conservatisms, such cost comparisons ensure that the severalfold price advantage shown for the soft technologies over their marginal competitors is not an artifact of arguable assumptions but a decisive conclusion.

As a further conservatism, these cost comparisons ignore all externalities. They value at zero all the soft technologies' advantages (Lovins 1977; Lovins & Lovins 1980; Sant 1981; SERI 1981) for more and better jobs, less inflation, relief of oil import dependence, less risk and environmental damage (including the virtual elimination of acid rain), better national security and resilience, reduced nuclear proliferation, greater equity and political attractiveness, enhanced Third World development, reduced risk of technical failure, and—last but not least—lower risk of climatic change. Avoiding these social costs is merely an added benefit.

4.2.3. NATIONAL, REGIONAL, AND GLOBAL ADEQUACY

As mentioned in Section 4.2.1, detailed assessments of proven, cost-effective renewable sources in a wide range of countries have shown they are capable of meeting essentially all long-term energy needs at an economically efficient level of energy productivity. We shall now explore in greater detail how the findings of some of these national studies can be applied on a wider scale.

The countries in which indigenous researchers have examined the economically attractive renewable potential in greatest detail are shown in Table 4.2. (Japan is shown separately because, while a good renewable assessment is available [Tsuchiya 1980, 1981], a correspondingly thorough demand assessment is not yet complete at this writing [September 1981].)

Table 4.2 *Potential renewable supply calculated in selected national studies*

Country and source	Renewables (primary EJ/y) and % of study's demand[a]	
	2000	2025 or 2030
UK (Olivier et al. 1981)	0.891 = 14%	3.068 = 55%
FRG (Krause et al. 1980)	1.301 = 15%	2.681 = 44%
France (Groupe de Bellevue 1978)	1.848 = 28%	4.422 = 67%
Denmark (Meyer et al.)	0.220 = 44%	0.332 = 83%+
Sweden (MALTE 1977; Johansson & Steen 1978)	0.972 = 81% (in 1990)	2.100 = 100% (in 2015)
US (SERI 1981; Sørensen 1980)	13.0 – 23.7 = 19 – 35%[b]	42.6 = 100%
Total for countries listed	18.2 – 28.9	55.2
Japan (Tsuchiya 1980)	6.17 (ca. 2010)	6.17 (ca. 2010)

a. Percentage of primary demand, calculated by each study for the stated year, that is met by calculated renewable contribution. Conservatism of demand estimates varies widely, as noted below. Supply is matched to end-use structure to avoid supplying more energy of a given kind than can be used. Supply is generally time-constrained, not asymptotic. 1 EJ/y = 0.0317 TW.
b. Averaging SERI demand range (65.4–69.5 EJ/y); depends on car efficiency.

To express the imputed long-term energy demands for these diverse countries on a more consistent basis, we next compare the national studies' demand estimates with demands recalculated using the method of Table 2.19: that is, assuming a 2030:1975 GDP ratio of 2.41 in Europe or 2.5 in the US, a relative energy service intensity per unit of GDP of 0.65 in 2030, and a relative energy intensity per unit of energy service of 0.26 in 2030. The results, shown in Table 4.3, reveal the expected wide variation in the degree of detail and technical conservatism reflected in the national demand estimates as measured against this common benchmark. While this comparison is only indicative, owing to compositional differences between countries, most of the demand estimates shown clearly reflect an uneconomic level of inefficiency.

Table 4.3 *Primary demand estimates from selected national studies, compared with calculations based on the FRG/Colombo & Bernadini results*

Country	EJ/y (kW/cap) in 1975		EJ/y in 2000		EJ/y (kW/cap) in 2025-2030	
			study	recalculated	national study	recalculated
UK[a]	9.3	(5.3)	6.48	6.84	5.26 (2.87)	3.79 (2.07)
FRG	9.7	(5.0)	8.73	7.14	6.07 (3.85)	3.95 (2.50)
France	6.1	(3.7)	ca.6.6	4.49	6.6 (3.5)	2.48 (1.31)
Denmark	0.8	(4.9)	0.5	0.59	0.4 (2.4)	0.33 (1.97)
Sweden	1.6	(6.2)	1.2[b]	1.18[b]	2.1[b] (8.1)	0.65[b] (2.51)
US	75.0	(11.2)	65–69	54.6	44.9 (5.66)	31.73 (4.00)
total	102.4		89–93	74.8	65.3	42.93

a. All UK figures include feedstocks; 1976 data used in place of 1975.
b. Data for 1990 and 2015 respectively.

Combining these demand estimates, both from the national studies and from our recalculations, with the studies' renewable supply estimates (Table 4.2), and adding a column for recalculated demand using our "present technical limit" coefficient, yields the renewable supply fractions of total demand shown in Table 4.4.

Thus completely renewable supply is difficult (though not impossible, as noted below) to achieve in the most unfavorable cases—the UK and FRG—with the 2.4-fold assumed increase in real GDP unless at the same time the "present technical limit" efficiency levels are assumed. The last column on the right shows that this combination brings renewable production into surplus. Table 4.4 also clearly shows the importance of assessing the degree of conservatism in end-use efficiency calculations before accepting a small projected solar fraction at face value.

A more detailed look at the reason for the initial renewable supply problem in the UK and FRG reveals that the key problem is not, as expected, the supply of vehicular liquid fuels

Table 4.4 *Potential renewable supply fractions for selected countries*

Country	National studies' renewable supply as % of their demand		Renewable supply as % of recalculated demand			Renewable surplus (deficit), EJ/y[d]	
	2000	2025–30	2000	2030	"t.l."	2030	"tech. lim."
UK[a]	14	55	13	81	153	(0.7)	1.1
FRG	15	44	18	68	128	(1.3)	0.6
France	ca.28	67	41	178	e	0.5	e
Denmark	44	83+	37	101	e	>0	e
Sweden[b]	81	100	82	323	e	1.45	e
US[a]	19–35	100	24–43	134	e	10.9	e
Japan[c]	—	(ca.100)	60	109	206	0.5	3.2

a. Includes feedstocks.
b. Values for 1990 and 2010.
c. Renewable supply nominally for ca. 2010; includes 3% geothermal and wavepower; GDP 2030/1975 = 2.41. No calculation specifically for 2000.
d. Assuming demand recalculated as above (relative energy service intensity of GDP = 0.65 "2030," 0.50 "technical limit"; relative technical efficiency = 0.26 "2030," 0.18 "technical limit"). Japanese recalculated demand is approximate.
e. Cases not calculated because 2030 supply was unproblematic.

(Krause et al. 1980; Olivier et al. 1981; Tsuchiya 1981) but rather of high-temperature process heat, particularly for the assumed large steel industry. The steel industry alone accounts for such a demand (using 2030 coefficients and assuming 2.4-fold GDP growth) of about 0.5 EJ/y in Japan and, with the cement industry and a few other high-temperature processes, also in the FRG. After improvements have been made in materials policy (Section 2.3.4; Lovins 1978a), the following options remain:

1. Fundamental change in the nature of steel markets. Eketorp, for example, believes (personal communication 1980) that most of the long-term market—especially as such major uses as car manufacturing change to lighter materials—will be not in large-section bars, ingots, and plates, but rather in filaments and ribbons to be incorporated into composite materials. If this occurred, both the scale and the inherent energy losses of steel fabricating machinery would drop dramatically, nearly to Second Law limits, using processes already developed.

2. Reduction of iron ore by charcoal. Recent Swedish experiments with modern versions of this old process show two advantages: the steel is of higher quality (because the wood contains no

sulfur), and methanol can be coproduced at zero energy cost—even exothermically—for the transport sector, or economically attractive fuel gas for industry.

3. Renewable energy supply using types or quantities of systems not assumed in the national studies. For example, Battelle and IIASA studies a few years ago showed considerable promise for central solar-thermal-electric conversion ("power towers") in Central Europe, and based on those studies, Caputo (1980) states that high-temperature process heat can be reasonably supplied there at a price equivalent to about $30–35/bbl oil ($5–6/GJ) in the near term and half that with advanced technologies. Either price is attractive at the margin. Alternatively, electric steelmaking could be based on extra photovoltaics (or, in Japan for example, offshore windpower—an immense resource). It is too early to tell whether such approaches are the best way to run a long-term steel industry; certainly the efficiencies considered here would enable even Japanese and FRG domestic coal resources, if reserved for primary metallurgy, to last at least for centuries. But whether economically optimal or not, wholly renewable provision of high-temperature process heat, directly or electrically (or via hydrogen), is certainly feasible with present technology.[4] The question is not whether this can reasonably be done, but rather which of several options is cheapest.

4. Exchange of surplus high-grade renewable energy between nearby nations may be the most attractive option. Although little analysis of intraregional renewable supply/demand balances has been done, the several examples available suggest great scope for cost-effective interchange. Caputo (1981), for example, has found that a Western Europe vastly more prosperous than today could meet virtually all its energy needs with renewables—and with high efficiencies, using soft technologies alone. Sørensen

---

4. Krause et al. (1980) assume FRG supply in 2030 of 780 PJ solar heat, 1465 PJ biomass, 348 PJ wind, and 88 PJ hydro, but no solar heat >300° C and no photovoltaics. Olivier et al. (1981) assume UK supply in 2025 of 1052 PJ solar heat, 1612 PJ biomass, 80 PJ wind (even though the potential is similar to or greater than that of the FRG), 38 PJ hydro, and 286 PJ geothermal, but again no photovoltaics and little high-temperature direct solar heat. Thus both have left considerable room for expanding the direct or indirect renewable supply of high-temperature process heat.

(1981) has reported preliminary results of a joint analysis among the Nordic countries, showing that they could trade surplus hydroelectricity (mainly from Norway) for surplus biomass liquids (mainly from Finland and Sweden) to mutual advantage, satisfying all the countries' energy needs at reduced total cost (and incidentally integrating these exchanges with others of food and fiber). Based on one scenario (out of four) which assumes a doubling of the 1980 GDP by 2030, and a simultaneous improvement of average specific energy intensity to 0.4× the 1980 values, the Nordic study has so far derived the illustrative supply/demand balances presented in Table 4.5, compared with the alternative demand levels recalculated using the methodology of Table 2.19.

Table 4.5 *Preliminary, approximate Nordic renewable supply/demand balances*

Country	Primary demand (EJ/y)			Renewables (EJ/y)		Surplus (EJ/y) potential renewables – recalculated demand
	1975	2030	2030 recalculated	potential	used	
Norway	0.77	0.42	0.31	0.86	0.73	0.55
Sweden	2.01	1.06	0.82	2.48	1.28	1.66
Denmark	0.72	0.41	0.29	0.50	0.50	0.17
Finland	0.93	0.65	0.38	0.75	0.75	0.37
Total	4.42	2.54	1.80	4.59	3.26	2.75

With Scandinavia as well as France likely to be net exporters of renewable electricity and liquids, and with the strong possibility (Caputo 1981) of surplus liquids, and perhaps hydrogen, from Southern Europe, completely renewable supply of heavy industry in the UK and the FRG evidently would not require the care or sacrifice the flexibility that our earlier discussion implied. North America would enjoy a similar surplus of high-grade renewables.

This intraregional example raises the question of balance *between* regions. Another analysis by Sørensen (1979) has already considered this question by calculating the balance of renewable energy flows—matched, like the Nordic exercise, to end-use demand categories—among seven regions of the world. For illustration and as a powerful conservatism, he assumed a population of 10 billion, and a worldwide level of material affluence twice that of Scandinavia today. He liberally supposed this to entail a per capita energy budget of 4 kW, comprising 1 kW of vehicular liquids, 1 kW of electricity + mechanical work + high-temperature process heat, and 2 kW of low- and medium-temperature heat (split 1.5/0.5 at high and 0.5/1.5 at low latitudes respectively, with medium-temperature heat including active cooling). Despite this very inefficient use, leading to global energy use of 40 TW, self-sufficiency in renewable energy (albeit not all soft technologies) could be reasonably achieved in each region in all categories, except for a modest deficit in Asia made up by exports of biofuels and hydrogen from regions with corresponding surpluses. Obviously, with more realistic efficiency assumptions, no imbalances would arise between regions, and the renewable supply system would be far more attractive. Its main technical assumptions were:

- net solar heat supplied 20 TW, or 0.2% of mean insolation (land-use and structural constraints were taken into account);
- 25% of end-use energy (probably about 1.3–2× the economically justifiable fraction and at least 5× the economically justifiable level) was supplied as electricity—1.07 TW from hydro at all scales, 0.2 TW wind, 0.3 TW geothermal, and 8.8 TW photovoltaics; and
- biomass, including some mariculture, yielded 9.7 TW.

These supplies are generally (except for hydro) a small fraction of technical potential and a very small fraction of the gross resource base.

An illuminating scoping calculation by T.B. Taylor (1979) further confirms the reasonableness of assuming indigenous renewable potential in large areas that have not been specifically analyzed. He noted that >99% of the world's people live in a solar regime between that of Stockholm and that of Lake Rudolph (Kenya), which differ by 2.5× in mean insolation and by about 24× in seasonal fluctuation ratio on a horizontal surface (much less with optimal elevation angle). By showing that his conceptual designs for low-cost solar systems—based on salt-gradient and ice ponds and on air-inflated plastic-film collectors, and now being successfully checked in town-scale experiments—are likely to achieve very attractive cost goals in both of these extreme climates, Taylor made a powerful *a fortiori* case for intermediate climates (bolstered by his case study of nine developing countries [ibid.:App. II]). His design philosophy appears compatible with low end-use efficiency, hence with high demands—a conservatism.

Caputo (1980) and IIASA colleagues have also assessed the technical potential of renewables, with some attention to their regional balance. They found a global potential, fairly well distributed, of about 2 TW of wind, 10 TW of biomass, 0–5 TW of photovoltaics, and 3–17 TW of on-site solar heat collection. This total of 15–34 TW from soft technologies (not counting an additional central-electric solar potential of 57–244 TW) provides a large safety margin by comparison with our calculated global demand (Section 2.4) of <4–8 TW total. This confirms the implication of Sørensen's (1979) 40-TW renewable scenario: at total world demand levels of a few to ten TW, as we conservatively calculate to be economically optimal for 8 billion highly affluent people, the problem is not likely to be providing enough soft-technology supply, but rather selecting the most convenient forms of it from an embarrassingly wide range of attractive possibilities.

So far this discussion has concerned only whether *all* of projected energy demand in an exceedingly wealthy and industrialized world could be economically met by a restricted list of presently available renewable sources. We have suggested that this is very probably the case, even in the least favorable countries, although as a matter of convenience it may be simpler to make good minor national deficiencies by energy trade than by a rigidly enforced autarky. But we have thereby shown far more than we need to show. The whole argument that energy supply can be *completely* renewable is much stronger than is necessary for our consideration of climatic risks; for that thesis does *not* depend mainly on assuming any particular future role for renewable energy sources. The time-buying benefits of burning fossil fuels more slowly—stretching the fuels we already have and delaying their CO_2 contribution to the atmosphere—*spring largely from our analysis of ways to raise energy productivity, not from our assumptions about displacing fossil fuels with renewable sources.*

Moreover, if *any* combination of efficiency improvements and renewables will, as we argue, succeed in gradually reducing the total rate of burning fossil fuel, then the integrated fossil-fuel burn, hence the CO_2 level in 50–100 years, will depend mainly on how much fuel is burned *early* in this replacement process while the rate of burning fossil fuel is still relatively high. In other words, our conclusions about long-term climatic risk are sensitive mainly to efficiency-raising policy actions taken in the *near* term, when events are more accurately foreseeable and analyzable—not in the long term, when they are more hazy. This dominance of the long-term cumulative CO_2 contribution by near-term events is a consequence of our finding that the rate of burning fossil fuel should, with an economically efficient energy policy, decline, rather than rise. This is the opposite of the usual finding—when indefinite, usually exponential, growth in the burn rate is assumed—that CO_2 levels will rise most rapidly in the long term. Thus our entire line of argument about the long-term potential of renewables is an independent reinforcement to our basic findings about energy efficiency. Once the rate of burning fossil fuel has been reduced to a few TW, the rate of adding CO_2 to the atmosphere is so slow that we shall probably have centuries, not just a few years, to worry about it—as we shall show further in Section 7.2.

5
By What Policy Instruments?

5.1. Strategies for implementation: conflicting philosophies

We have already described—in some detail in our FRG case study (Section 2.3) and more broadly elsewhere—the economic incentives favoring a "soft" energy path based on efficiency and appropriate renewables. Probably of equal importance in the way societies actually make energy decisions are such social advantages of a soft path as reducing vulnerability to all kinds of disruptions (Lovins & Lovins 1981); allocating energy and its social costs to the same people at the same time rather than inequitably to different people at opposite ends of the distribution system; avoiding the need for centrist or technocratic decision-making; and protecting individual choice and local autonomy. So powerful is this convergence between political and economic logic that many observers are coming to suspect that the question is not so much whether as when and how smoothly the soft path will be realized.

There are, however, two broad concepts of how to design and implement energy policy. One school, identified in American tradition with followers of Alexander Hamilton, holds that these complex technical matters must be decided and enforced by a central elte. Another school, more identified with Jeffersonians (and free-market economists), holds that people are, on the whole, pretty smart, and that if they see the energy problem as *their* problem, and have incentive and opportunity, they will largely solve it themselves. Under the former view, energy policy requires elaborate central management which, under the latter view, it cannot tolerate.

This dispute is not technical, and it cannot be settled on technical merits. But at least in the Western democracies, the large-scale experiment that has been conducted since 1973 has soundly vindicated the Jeffersonians. The energy problem is indeed being rapidly solved, but from the bottom up, not from the top down; central governments may be the last to know. The energy problem is made up of such a gigantic number of little pieces, scattered throughout a diverse society, that central management is very likely to be more part of the problem than part of the solution.

Governmental action is seldom neutral; it can either help or hinder the transition away from fossil fuels, and can direct it towards or away from a climatically benign energy system. In the Western democracies, two prominent instruments for helping a soft path to implement itself come to mind: providing correct price signals and removing market imperfections ("institutional barriers"). We next sketch some elements of each approach.

5.2. Price signals

There is no true free market in energy or anything else. Strictly speaking, there may never be one, because the conditions under which a theoretical free market reaches an efficient equilibrium are quite onerous: perfect information about the future, perfect competition, no monopoly or monopsony (domination by one buyer), no unemployment or underemployment of resources, and so forth.[1] But even if these idealized conditions are not attained, at least one kind of information that greatly aids the efficient short-term allocation of limited resources is to have prices that tell the truth: prices that correctly signal relative abundance or scarcity, without distortion by subsidy or by promotional tariff structures. In particular, incremental consumption should as nearly as possible attract incremental cost, so that those who want more and thereby force society to incur higher marginal costs will have a way of knowing what their extra demand is

---

1. Welfare economics rests on odd assumptions too (Junger 1976; Lovins 1977a).

costing and hence of deciding how much is enough.

Unfortunately, most countries have substantial subsidies, in many subtle forms and often unevenly distributed, to make energy look cheaper than it really is. The extent of these subsidies in the FRG is probably smaller than in many other European countries, but is not really known. Although some countries have introduced special subsidies to renewable sources, it is better economics to desubsidize all the rest instead and let all options compete fairly at the margin. The most defensible residual subsidies will be those that tend to improve market structure, enhance competition, compensate for uninternalized social costs of alternative investments, and not be self-perpetuating. Such subsidies should be few indeed, because any subsidy to supply—even to benign sources—leads to underinvestment in energy productivity, which is notably free of externalities, risks, and inequities. Indeed, there is a case to be made (Lovins 1978d) for moving gradually and fairly via a depletable-fuel tax (charged on energy content as the fuels come out of the ground or into the country) toward pricing all nonrenewable fuels (and other resources) at their long-run replacement cost—something OPEC has been moving us towards.

Higher energy prices, no matter how they are attained, arise two problems: efficient price levels may be unfair in a society whose assets are unevenly distributed; and many people do not have enough capital to invest in measures to relieve them of the high prices. In particular, nearly all the efficiency improvements described earlier are very attractive at present energy prices, and do not require marginal prices to elicit them; yet it is often said that if prices rise enough, people will somehow adjust. If that adjustment is to be achieved without great hardship, capital must be as easily and cheaply available for efficiency as for supply investments, even if the former investments are made by millions of individuals rather than by a few utilities. We have proposed elsewhere (Lovins 1981b,c) a mechanism—the elements of which are already in use in the US—for loaning capital from the supply industry to end-users whenever that will result in lower-cost provision of energy services. Such loans can both relieve the capital burden on consumers and offer great financial benefits to the lenders. (Electric utilities representing >40% of US generating capacity are now giving or preparing loans for efficiency improvements, though seldom on terms as mutually advantageous as we propose.) The mechanism we propose has the further attraction that the "investment balancing test" used to decide whether capital should be allocated to new supply or to end-users' alternative investments would compare all alternatives with the replacement cost of energy represented by the proposed new plant. Most of the capital going into the national energy system would thus be allocated *as if* energy were already priced at marginal cost whether it is actually so priced or not. Thus a least-cost allocation can be achieved without having to pay very high prices first.

5.3. Market imperfections

Correct price signals are useless if people cannot respond to them.[2] Our societies have accreted thick layers of customs, rules, and institutions left over from the cheap oil era, and which now inhibit alternatives to it. Already mentioned are inequitable access to capital and to information: these two are often the greatest barriers to efficient choice. Other common problems include obsolete building codes or outmoded lending or land-use-planning regulations (many of which actually outlaw efficient buildings or various renewable sources); restrictive or anticompetitive utility practices; some architectural fee structures which inhibit efficient design; and, pervasively, split incentives between builders and buyers or landlords and tenants. (Why should the landlord stuff up the cracks around the windows if tenants pay the heating bills? Why should tenants fix them if it's not their building?) There are solutions to all these problems (Lovins & Lovins 1980:Ch.8), but they must be locally tailored and are undoubtedly messy. These are not easy problems: only easier than *not* solving them.

The institutional rigidities and momenta that make any effec-

---

2. Consumer psychology is also crucial, especially in realizing that light and mechanical work take relatively little energy but heating takes a great deal. A kW-h can run a 100-W light bulb for 10 hours, or lift a ton more than 300 m straight up, but it can heat by 30 C° less than 30 liters of water—only a few minutes' shower.

tive government action hard to start, harder to deflect, and hardest of all to stop are more or less the same all over the world. A pragmatic approach to energy policy may tempt one to involve government, especially central government, as little as possible, and to rely more on local adaptation. But this is not the place to offer advice specifically suited to the conditions of any particular country. We can only suggest that international experience in these matters offers a rich menu of possibilities, many of which may be translatable to other cultural and political settings. We have also catalogued elsewhere (ibid.:135–140) some special ways in which removal of institutional barriers, supply of psychological leadership, and selective, short-term help to grassroots efforts can help to overcome the unique problems of implementing energy efficiency and renewables in developing countries.

5.4. Accelerating retrofit or turnover of capital stocks

Special mechanisms are available to local or central governments which wish to fix or replace energy-inefficient or fossil-fuel-based capital stocks more quickly than their normal rate of repair and attrition, so as to save oil or cut CO_2 emissions faster. Indigenous programs have successfully retrofitted major capital stocks in a few years to a decade in such cases as the Swedish conversions to right-hand driving, 230-V electricity, and (currently) district heating; the Dutch conversion to Groningen gas; the British conversions to North Sea gas, decimal coinage, indoor toilets, and smokeless fuels; and, perhaps most interesting, the conversion of metropolitan Toronto and Montréal from 25- to 60-Hz electricity. The Canadian program used resourceful, improvisatory technicians in fleets of specially equipped vans, who retrofitted one neighborhood at a time after ample public education. One van contained a machine shop which could rewind motors and rebuild controls; another offered a choice of many styles of clocks at the new frequency in exchange for old ones; and so on. Presently the job was done and the vans moved on to the next area. Similar methods could be used—with substantial cost savings—for efficiency/solar retrofits of buildings.

It is also important to consider accelerating the turnover of inefficient consumer durables. Some US utilities are considering, for example, buying inefficient refrigerators and scrapping them, because that would be cheaper than building new power stations. The Tennessee Valley Authority has been designing a system whereby electricity-saving measures would be eligible, just like self-generated electricity, for buyback. (Under the Public Utility Regulatory Policies Act of 1978, US utilities must buy back consumers' surplus power at "avoided cost," which is supposed to mean as much money as the transaction saves the utility.) TVA would therefore give, for example, a voucher which could be used to help buy a new refrigerator, and whose value would represent TVA's "avoided cost" from the resulting electricity savings. Utilities in Texas and Minnesota already repay, in certain circumstances, electricity-saving investments with vouchers applicable to utility bills; and in a New Jersey experiment, third-party efficiency investments are providing saved energy to the investor, in much the same way that decrements of air pollution have become a marketable commodity.

Similar mechanisms are undoubtedly needed to relieve the capital burden of buying a more efficient car. Although there is ample economic incentive, even in the FRG (see Section 2.3.2 above), for changing cars, many people simply do not have the money. In the US this is a serious problem because the gas-guzzlers have very low trade-in value, and are therefore trickling down to the poorest people, who cannot afford to run them *or* replace them. The residence time of these inefficient cars in the stock is therefore increasing when it should be decreasing. This suggests the following illustrative proposals:

- Rather than spending $20 billion (plus perhaps $68 billion more later) to subsidize synfuel plants, the US could save more oil faster by using part of the same money to pay at least half the cost of buying people a diesel Rabbit or equivalent, providing they would scrap their Brontomobile to get it off the road. (It cannot just be traded in, because then someone else might drive it; it must be recycled and a death certificate provided for it.)

- Alternatively, the US could get a five-year average payback compared to synfuels by paying people at least $200 for every mile per gallon by which a new car improves on a scrapped one. (People who scrap a gas guzzler and do not replace it should get a bounty for it.)
- Instead of merely redirecting synfuel subsidies into better buys, as in the two preceding examples, it would be better yet to abolish the subsidies and use a free-market solution. If the US car industry spent as implausibly large a sum as $100 billion extra,[3] probably enough to rebuild Detroit, on prompt retooling to make, say, 60-mi/US-gal cars (4 ℓ/100 km)—considerably worse than the best German and Japanese prototypes today—and then spread that marginal retooling cost over a new US fleet of cars and light trucks, then the extra cost, averaging $770 per vehicle, would be recovered by the buyers, at early-1981 US gasoline prices, in only 14 *months*. The trouble, of course, is that Detroit does not have the money (or, perhaps, the ability to move that fast anyhow). There may be some way—without antitrust complications—to persuade the oil companies, which do have the money, that they can drill for oil more cheaply (<$7/bbl), and be much surer of finding it, by lending Detroit money for retooling than by sending oil rigs to the ends of the earth.

These examples are of course only slightly whimsical illustrations, but they have a serious point. One should be neutral among all ways of providing an energy service, whether by raising energy productivity or by increasing supply. If the former is cheaper, one should seek ways to encourage the traditional, well-established financial mechanisms that have funded the latter to fund the former instead. This can only help to reduce inflation and interest rates, make abundant jobs,[4] and free capital for more productive uses.

3. Detroit plans to spend almost $50 billion on retooling in the 1980s anyhow (von Hippel 1980).
4. Power stations, in contrast, are so extraordinarily capital-intensive that each GWe built in the US destroys about 4000 net jobs by starving other sectors for capital (Hannon 1976). Careful and detailed US case-studies (Buchsbaum et al. 1979; Rodberg 1978; Schachter 1979) have shown that solar/efficiency programs provide several times as many jobs per kW as power-station investments, but better distributed by location and

While one cannot make cars energy-efficient in exactly the same way as buildings, the practical problem in these two instances is much the same: people who live in buildings or who own cars have an economic incentive to make them more efficient (by retrofitting the buildings and replacing the cars) but may lack the opportunity to do so. Appropriate capital transfer mechanisms can go far to accomplish both. The benefits are enormous. In the US, for example, the switch to a 60-mi/US-gal (4 ℓ/100-km) fleet of cars would save nearly four million barrels of oil per day: two-thirds of the entire 1980 net rate of US oil imports, greater than imports from the Gulf, two and a half Alaskan North Slopes, or 80 big synfuel plants. (Similar action with light trucks would save a further one and a half million barrels per day, or about one North Slope.) Basic weatherization of US buildings would save at least two and a half million barrels per day by 1990 at a cost of about $6–7/bbl (DM 0.08–0.16/$\ell$) (SERI 1981). Thus just the two biggest oil-saving measures alone, pursued just over the next decade to a level well short of what is technically feasible or economically worthwhile, would together more than eliminate all US oil imports—before a synfuel plant or power plant ordered today could deliver any energy whatever, and at about a tenth of their cost. The US oil saving would also reduce the global rate of CO_2 emission by about 5%.

Likewise in the FRG, a shift of the car fleet from 1973 efficiency levels (10.6 ℓ/100 km) to 4 ℓ/100 km would have saved, at 1973 car and driving levels, over 14 billion liters per year of gasoline, worth at early-1981 prices some $9 billion per year, or an average of $530 per car per year. That sort of cash flow, which only increases with oil prices, in the hands of German consumers rather than of OPEC, could have appreciable multiplier effects, and the investment that produces it is inflation-proof. Since buildings used over three times as much fuel as cars did in the FRG in 1973, there is presumably an even greater scope for relatively fast savings there—if only the capital investment were directed into a least-cost energy strategy.

---

occupation. Many progressive US labor unions are now supporting a soft energy path. Similar evidence and trends are emerging in Canada (Brooks 1981), Australia, and elsewhere.

6
How Quickly Can These Things Be Done?

The technical and economic ability of present efficiency improvements and renewable sources to provide essentially all the energy services that humankind might foreseeably need, continuously and indefinitely, is firmly established in the minds of those analysts who have closely studied the state of the art. But actual deployment of the required millions, even billions, of devices is a different matter. We therefore consider in this chapter how quickly, given at least non-hostile policies, this deployment might reasonably occur, drawing on recent empirical evidence of what seems to work. While the forces which determine deployment rates are largely political, and their future course is thus somewhat conjectural, we shall adduce arguments that the "soft energy path" whose key technical elements we have described (Lovins 1977) is likely in practice to be realized faster than an equivalent nonrenewable-based policy could be, provided that strong obstacles are not deliberately put in the way.

6.1. Rate and magnitude problems

The central practical question about any energy policy, whether for displacing oil or for minimizing the climatic effects of burning fossil fuels, is how fast it can successfully achieve particular levels of supply or equivalent efficiency improvement. The formidable rate-and-magnitude constraints of new conventional energy supply, especially from technically or socially complex devices with inherently long lead times, are seldom as well appreciated as are those long lead times themselves. Illustrative calculations about the potential role of nuclear power—applicable, *mutatis mutandis,* to coal-fired or hypothetical fusion power stations, synfuel plants, or any other system that comes in blocks producing gigawatts, costing billions of dollars, and taking about a decade to build—are summarized in Table 6.1, with details given elsewhere (Lovins 1978; Lovins & Lovins 1980). The combination of long lead time and (in a national perspective) limited unit capacity results in an inherently slow and modest contribution even from very ambitious programs, even assuming away all financial, economic, and political constraints. This is the reason that scenarios published by both IIASA and the Institute for Energy Analysis at Oak Ridge (Weinberg et al. 1979) do not, on careful examination, show that even a massive nuclear program could much delay attainment of high CO_2 levels. Even the most generous central-electric programs, as the last column in Table 6.1 shows, are too little, too late, against the background of fossil fuel use rising more slowly but from an enormous base value.

Clearly, then, the only important variable in CO_2 timing—as we also found from Table 1.1 and Figure 1.1—is efficiency improvements. In the mid-1970s it was generally assumed that these would be very slow; but even analysts who thought they might be fast have been startled by their actual speed. In the nine EEC countries from 1973 to 1978, for example, primary E/GDP ratio decreased by about 8% while primary energy consumption increased 0.42%—a ratio of about 19, implying that about 95% of all economic growth was fueled by energy savings and only 5% by all net supply expansions combined (St. Geours 1979: Ann. 2:p.1, cf. p.2). The E/GDP elasticity in that period was negative in the UK, The Netherlands, Belgium, and Luxembourg; the highest E/GDP elasticities were 0.4 (Denmark) and 0.3 (FRG). Later figures were even better as real price increases started to bite. In the US, even more strikingly, the ratio of energy savings to new supply was 2.5:1 for 1973–78, >50:1 in 1979, and almost infinite in 1980, when real GNP was essentially flat (within about 0.1%, well inside the statistical noise) while primary energy use

Table 6.1 *Some illustrative rate-and-magnitude calculations for displacing fossil fuels by central electrification*

	Assumptions					Results			
region	target date D	demand[a]	J(f)/ J(e)[b]	nucl. goal @ date D	constr. time, y	days/ GWe[c]	cost[d]	final GWe[e]	fossil-fuel[a] use @ date D
US	2000	1978 PT + 1.72%/y[f]	2P	(PT in 2000)/4	12	5	>0.75 ×NPDI		
France	2000	1975 PT +3.8%/y					FFr 6 ×10^{11}	104	1975 O&G + 70%
Japan	2000	PT = 41 EJ/y[g]	2D		6	20	¥10^{14}	274	1978 O&G+ 73%
OECD Europe	2000	1975 DT +1%/y; 1975 DE +2%/y					$3 × 10^{11}	210	1975 + 13%
same	2025	same					$10^{12}	614	1975 + 30%
same	2000	1975 DT +2%/y; 1975 DE +4.4%/y					$6 × 10^{11}	384	1975 + 33%
same	2025	same					$3 × 10^{12}	1890	1975 + 92%
world	2000		2P	(1970 O&G)/2	8	3.5	$3 × 10^{12}	> 1600	

a. P = primary; D = delivered; T = total; E = electrical; O&G = oil & gas.
b. J of fuel assumed to be displaced by each J of electricity supplied.
c. Average ordering interval (days) between 1-GWe plants to attain goal shown.
d. Approximate minimum cost of new stations and associated marginal grid, in 1976 currencies (prices from Lovins [1977,1978b,c,1979]). In US column, NPDI is 1977–2000 integral of net private domestic investment (assumed 7% of GNP with GNP growing 3.5%/y), but cost excludes stations not yet operating by 2000.
e. Assuming 62% capacity factor (US reactors >800 MWe averaged 54% to VI.80).
f. This yields 120 EJ/y in 2000, the IX.79 median USDOE and VI.80 "preferred" Edison Electric Institute forecasts.
g. Assuming trebled GNP (below 1978 MITI forecasts) and conservation 1/3 better than MITI's 1978 "maximum possible" projection.

fell 3.4%. There is some evidence that the portion of this saving that arose from improved technical coefficients was almost entirely "low-cost/no-cost" or "good-housekeeping" measures not requiring significant investment. If gradually diminishing returns on efficiency investments were to counterbalance gradual improvement in today's minimalist policies (notably by removal of market imperfections) to produce a more or less linear rate of long-term implementation, the 1974–78 EEC record (8% saving) would imply an 80% (five-fold) saving in 50 years without anyone's really noticing, and the US 1980 rate would imply an 84% (6.1-fold) saving! This makes the three-fold UK efficiency improvement described by Leach et al. (1979) look almost staid: indeed, that trebled efficiency, and more, would be achieved in 50 years if the far from impressive rate observed in the UK in recent years would merely continue unabated.

To put it another way: the EEC's 8% primary energy saving for 1973–78 amounted to some 3.1 EJ/y primary, equivalent to about 2.36 EJ/y of delivered energy. But that much delivered energy would have been supplied (at the actual 1978 EEC capacity factor and grid losses) by 145 GWe of additional installed nuclear capacity, which is 10.9 times as much as the EEC actually installed between 1973 and 1978 at considerable economic and political cost. Even in France during 1973–79, the world's most aggressive nuclear program provided at most 31% as much new energy as a not-very-serious program to save energy. Likewise, the US between 1973 and 1978 actually got twice as much energy, twice as fast, from energy savings as synthetic-fuel advocates claim they can provide at ten times the cost (if they are given $88 billion to get started with). Since the mid-1970s, *efficiency has been by far the fastest-growing part of world energy supplies*.

Local examples are often even more impressive. Nova Scotia homeowners used no-strings grants of C$350 to weatherize half the housing stock in a year. In Fitchburg, Massachusetts, a pilot program of door-to-door citizen action with no grants weatherized the worst fifth of the housing stock in a couple of months. Such actions add up. Heat losses in oil-heated FRG single-family dwellings fell by 20% between 1973 and 1979 (Schipper, personal communication 1980). Denmark cut total direct-fuel use by 20% just in 1979–80 (Nørgard 1981). Japan has had seven years of zero energy growth with 4%/y average GNP growth (and 7.2%-

renewable primary supply—nearly the official forecast of 8% in 1990 [Tsuchiya 1981]). Millions of individual actions in the marketplace—people seeking to save energy to save money—are outpacing the centrally planned supply programs by tens or hundreds to one. Though supply planners are often reluctant to rest their confidence on those uncontrolled individual actions, precisely the same mechanisms are at work which have always been invoked as the rationale for forecasting *growth* in demand. The countless individual market decisions which make up national demand are simply responding to an altered set of signals.

6.2. Rate constraints and comparisons

Many of the generic constraints to increasing energy supplies are well-known. Among the most prominent in many countries currently are finance (Lovins 1981b,c), macroeconomic problems, difficulties in finding acceptable sites, and the sheer logistical problems of organizing vast construction projects with appreciable technical risks, a sometimes immature technical base, and a fluid regulatory environment (Komanoff 1981).

Added to these uncertainties are those of demand, on which investors count to safeguard their commitment of billion-dollar blocks of capital. If demand is even a fraction as sensitive to price as our earlier arguments suggest, then higher price—which an energy company must charge to recover the higher marginal cost of a new plant—may reduce the *amount* of energy sold more than the company can compensate by charging more money for each *unit* of energy sold. In this case, new investments will require higher revenues but actually produce lower revenues (negative price elasticity of revenue), leading quickly to insolvency. Whether or not this turns out to be true globally (it certainly is in the UK and some other countries), the combination of extreme capital intensity and long lead time makes cash flow in such enterprises—especially for electric utilities—inherently unstable (Lovins 1981b,c); so if utilities keep building plants, they will go bankrupt, regardless of whether they are regulated perfectly, badly or not at all.

These formidable *fiscal* problems must be considered jointly with the underlying *economic* problem that many forms of energy—especially electricity—no longer have a marginal market. In the industrialized countries, at least, they cannot compete with efficiency improvements (ibid.). Sant, for example, showed (1979) that at rolled-in 1978 US prices, about 43% of the electricity sold was uncompetitive. At today's prices, the fraction is larger; at marginal prices, it may be large enough to include all the thermally generated electricity in the United States, leading to the unprecedented prospect of having to write off, over the next decade or two, over $100 billion net worth of uncompetitive thermal generating plants.

We mention these problems because the reality of the opportunities we have described is not merely an academic matter; it is the crux of decisions, in both the private and public sectors, on which depend a substantial fraction of total national investment and fixed assets. Overinvestment in supply to meet a demand which does not arise—because consumers have wisely chosen a cheaper way to do the same tasks—has an unconscionable opportunity cost and can deprive an energy supply industry (or even the national treasury) of its financial room to maneuver: for example, into investments paying a higher and faster return (Lovins 1981c). Once the capital markets perceive that risk, as they have lately been doing in the US utility debt and equity markets, investors are no longer willing to expose their money to the risk of losing it—certainly at the return they are offered and often at any feasible rate of return. Lack of finance itself then becomes the most serious constraint on new energy supply, far more ineluctable than the problems of siting and regulation.

As forecasts of primary energy demand fall (Table 4.1), that does not mean that some planned increment of supply (so many power plants or synfuel plants or LNG terminals) can make a proportionately larger contribution. Quite often, on the contrary, the fractional contribution may decline. This is because primary demand growth in virtually all official forecasts is dominated by electrical growth, which is counted in terms of the primary fuel which is (or would be in an equivalent fossil-fueled plant) burned to raise steam for turbo-generators. This means the electricity itself is counted at roughly treble value. Reductions in energy forecasts are therefore three times as sensitive to reductions in the forecast demand for electricity as for direct fuels used to provide mobility or heat. Consistent with this peculiar sensitivity, most reductions in primary demand forecasts are in fact due largely to reduced prospects for electricity demand; and for technical reasons, only about half that demand in turn can be met with baseload plants. Thus, reasoning backwards from the demand forecasts to their supply components, lower primary demand does not leave room for unaltered baseload electric supply to provide a larger share of total energy supply, but instead often reduces the baseload electric component disproportionately. This is not the case with end-use-matched soft technologies, since they provide predominantly direct heat and vehicular liquid fuels corresponding to the end-use demand structure. Thus reductions in primary demand forecasts often permit a given amount of soft-technology supply to represent a correspondingly larger fraction of total energy supply.

6.3. Comparative rates: theory and practice

Several characteristics of efficiency improvements and, secondarily, of soft technologies suggest that they ought to be the fastest measures available, per unit of investment, for providing needed energy services. Most obviously, each unit takes days, weeks, or months to install, not a decade. Secondly, those units can diffuse rapidly, almost spontaneously, into a large consumer market—like digital watches, pocket calculators, and (in the US) citizen's-band radios—rather than requiring a tedious process of "technology delivery" to a narrow, specialized, and perhaps "dynamically conservative" utility market (like 1-GWe power plants or basic oxygen furnaces). This is a function of the relative understandability, marketability, and accessibility of soft technologies—of their comparative technical and managerial simplicity and the ease with which they can adapt to local conditions.

Technologies that can be designed, made, installed and used by a wide variety and a large number of actors can achieve deployment rates (in terms of total delivered energy) far beyond those predicted by classical market penetration theories. As a

Gedankenexperiment, let us imagine two sizes of wind machines: a unit of several MWe peak capacity, which can be bought for perhaps $1 million and installed by a heavy engineering contractor in a few months on a specially prepared utility site; and another of a few kWe, which can be bought by a farmer on the Great Plains of North America (or in North Germany) from any local auto parts or equipment store, brought home in a pickup truck, put up (with one helper and hand tools) in a day, then plugged into the household circuit and left alone with no maintenance for 20–30 years. (Both these kinds of wind machines are now entering the US market.) Most analysts would emphasize that it takes a thousand small machines to equal the gross output of one big one—actually less, because the small ones, being dispersed, are collectively less likely to be simultaneously becalmed (Kahn 1979; Sørensen 1979a). It may also be important that the small machines can be produced more than a thousand times as fast as the big ones, since they can be made in any vocational school shop, not only in elaborate aerospace facilities, and are probably cheaper per kW to boot. But what may be most important—and is hardly ever captured in this type of comparison—is that there are three orders of magnitude more farms than electric utility companies on the Great Plains, and a farmer can often act faster than a utility executive.

The third reason for suspecting that many small, simple things should be faster to do than a few big, complicated things is that the former are slowed down by diverse, temporary institutional barriers—passive solar by the need to retread architects and builders, microhydro by licensing problems, greenhouses by zoning rules—which are *largely independent of each other*, as opposed to the generic constraints on "hard technologies" (major-facility siting problems, large-project finance, etc.). Because of this independence, dozens of individually slowly-growing soft-path investments can add up, by strength of numbers, to very rapid total growth, rather than being held back by the same problems everywhere at once. The diversity of renewable and efficiency options is not only a good insurance policy against technical failure; it also helps to guard against specialized, unforeseen social problems in the course of implementation, always providing an alternative way to evade what problems do arise.

How do these theoretical considerations square with practical experience? Very well indeed. Let us pick a few examples from the United States, whose geographic and social diversity and pluralistic market process are a good match to the technical options' diversity. Efficiency improvements, as noted earlier, are moving ahead with extraordinary speed, unforeseen by any of the energy industries. This is causing many oil and electric companies financial hardship, though energy-saving and energy-managing services are Royal Dutch Shell's most profitable subsidiary business, and are proving an extremely high-return private investment: one such firm, started two years ago, is expecting $250 million turnover in 1983. Americans spent some $8.7 billion on energy-saving devices in 1980 and are expected to spend tens of billions of dollars per year by the mid-1980s (*Business Week* 1981); Sant et al. (1981) showed that $100 billion per year would be economically worthwhile.

How about renewables? The US, which already gets about 7% of its primary energy from renewables (mainly hydroelectricity), is now approaching its millionth solar building, of which half are passive, and half of those are retrofits (mainly greenhouses, plus some glazing of exterior masonry walls to form Trombe walls). Many of these were built on the basis of word-of-mouth or popular journal information, few from officially provided information. In the most solar-conscious areas, about 6–7% of all space heating is solar, and 25–100% of the new housing starts are passive solar designs. Nationally, about 15% of the contractors building tract houses, and virtually all purveyors of prefabricated and package-planned houses, are offering thermally efficient, passive solar designs. In New England, over 150 factories have recently switched from oil to wood, as have more than half the households in many rural (and some suburban) areas. Vermont in 1980 burned more wood than heating oil. Private wood-burning in the US has increased more than sixfold in the past few years, and the number of stove foundries has risen from a handful to more than 400. There are over 40 main wind-machine companies. Commercial "windfarms" are now competing on utility grids. Hawaii plans to get 9% of its electricity from wind by 1985. Well over 10 GWe of small hydro capacity is under reconstruction (mainly refurbishing old, abandoned dams). Permits were sought for a further 20 GWe of small hydro during 1979–80—twice the gross nuclear capacity ordered during 1975–80. Small-hydro development companies, and entrepreneurs who collect surplus electricity to sell back to utilities, are springing up in many areas. There are over a thousand retail outlets for Gasohol®, and most states have biomass fuel programs. Several geothermal industrial parks are under construction.

It is hard to find a part of the US that does not have its unique blend of renewable energy ferment. Many observers who travel the country remark that although these activities are concealed from governments' view by their dispersion, small individual scale, and diversity, they add up to a quiet energy revolution that is reshaping the American energy system—and some others (e.g., Tsuchiya 1981)—with unprecedented speed. The proof of success: collectively, renewables have contributed more new energy to the US since 1977 than all other supply sources (RTM 1981)—probably even more than the vast expansion of coal-mining that has so concerned climatologists.

6.4. Oil displacement and other short-term imperatives

Many countries share the Federal Republic's sense of urgency in relieving dependence on imported oil—a dependence growing rapidly more costly in money, security, freedom of action, and macroeconomic damage. Indeed, when told that replacing oil means switching to CO_2-releasing coal, they often remark fatalistically that the urgency of oil displacement today outweighs the more remote and speculative prospect of climatic changes in the next century. But this is, on our analysis, a false dilemma, because *the same policies* can simultaneously save oil *and* protect climate. For governments anxious to save money, oil, and climatic risk, we therefore abstract from a fuller treatment elsewhere (Lovins & Lovins 1980:esp. Ch. 7) a brief summary of the principles that ought logically to guide policy decisions in this area.

First and foremost, if, as Sant and SERI have shown for the

US case (1981), a least-cost energy strategy also has the incidental effect of rapidly eliminating oil imports—because oil cannot compete in cost with other available options, notably efficiency improvements—then one need not necessarily pursue an oil-saving policy as a supreme goal: a practice which in some countries has so far distorted market signals as to have the opposite of the desired effect.

If, secondly, one is deliberately channeling investments into those directions which will save oil (or, on the Sant hypothesis, provide the most energy services) *fastest and cheapest per dollar invested*, one should symmetrically compare all marginal investments to see which will do this, then start with the most cost-effective and work down the list. Failure to make this comparison has led many governments to sink scarce national resources in programs—especially central electrification schemes—which, as we have suggested here and elsewhere (Lovins & Lovins 1980), are the slowest and costliest known way to save oil, and which therefore *slow down* oil displacement by diverting investment from more effective measures.

Thirdly, one ought to pay the most attention to the main uses of oil today: direct heat (in buildings and industry) and transport. Only about a tenth of the world's oil consumption (a rapidly declining fraction)—6% of the 1978 FRG final oil use—generates electricity, and Table 3.1 suggests that at least the industrialized countries already have more electricity than they can use to economic advantage. In North America and Europe, therefore (and in Japan, though there one should put perhaps more emphasis on industry), the prescription for saving large amounts of oil is distressingly simple: stop living in sieves and stop driving Petropigs (see pp. 133–135 above).

The ineffectual nature of pursuing oil displacement in electric utilities in preference to these major opportunities is illustrated by Dr. Vince Taylor's calculation (1979) that if in 1975, *every* oil-fired power station (thermal or internal-combustion) in OECD had been instantaneously replaced by some other sort of power station, such as nuclear reactors, OECD oil consumption would have fallen by only 12%, and the imported fraction of that consumption would have fallen only from about 65% to about 60%

(compensated by greatly increased dependence on imported capital and uranium). The overnight substitution would have reduced France's 1975 oil consumption by 10%, the FRG's by 4%, and Britain's by 14%. (These changes are so small because a third of French electricity was from hydropower, and nearly two-thirds of British and German electricity was from coal.) The imported fraction of national oil consumption would have fallen by much *more* for the United States (7.3 percentage points) than for Japan (2.6), France (0.5), the FRG (0.3), or Britain (0.8). This shows that the scope for displacing utility oil imports depends not only on how much of the oil used by a country is imported, but also on how much of that oil is used in power stations (and particularly in baseload power stations). That is why US nuclear expansion, for example, has served mainly to displace coal, not oil. It is fallacious to equate a kilowatt-hour of electricity with 0.3 liters of oil—the amount that would have to be burned in a rather inefficient power station to make a kilowatt-hour of electricity. The two are not simply substitutable—not even technically, and certainly not economically: FRG electricity delivered at \$18/GJ now costs about three times as much per unit of heat contained as \$36/bbl OPEC oil at \$6.2/GJ.

It is commonly said that renewable energy sources will take a long time to develop and deploy; that energy savings are also slow and costly and are at best a partial answer; and that only large increments of conventional supply technologies can therefore be counted on in the near term. But failure to assess *comparative* rates of oil displacement runs the risk that, having dismissed renewables as slow, efficiency improvements as costly, and both as inadequate, one may choose an investment strategy that is simultaneously slow, costly, *and* inadequate. If speed is of the essence, speed must be assessed.

The very short lead times for getting efficiency improvements and soft technologies in place enable them to start displacing oil (and saving energy and money) immediately, not in 10 years. Further, decisions to deploy these technologies need not depend on whether, which, or how much indigenous fossil fuel a country has available. It is sometimes argued that an ultimately solar-based economy, though doubtless attractive for a country which (like the United States) has abundant transitional fuels, is out of the question for countries dependent from day to day on massive oil imports. But this argument asymmetrically begs the question of what fuels such fuel-short countries will use as a bridge to their proposed dependence on uranium or coal instead, for that shift too will take time: on our arguments, even longer. In other words, whether a country has fossil fuels of its own, or which ones, or how much of them, has nothing whatever to do with whether conventional or soft-path investments can displace that country's oil use faster and cheaper. It is this question that our review of both theory and practice has addressed.

If saving oil is a national priority, then the decision rule that should be applied—whether by government planners or by a free market—is to compare in costs, rates, risks, uncertainties, and side-effects *all* the ways of achieving that goal, and start with the best. Doctrinaire assumptions about these conclusions are no substitute for sound analysis. We are persuaded by the foregoing arguments, as was the Harvard Business School study (Stobaugh & Yergin 1979), that efficiency improvements are by far the most attractive immediate investment from all these points of view, and that appropriate and proven renewable resources come next. Next in priority of cost and difficulty are synthetic fuels; and last—costliest and slowest of all—are power stations. By never making that ranking or allowing a free market to do so, the principal nations of the world have taken those choices in reverse order, worst buys first. But the present strains in the energy system—stagnating demand, overinvested supply, devolution of decisions from supply industries to end-users—are a sign that economic priorities are at long last reasserting themselves.

7
Implications for Climatic Risk

7.1. The cost of failure

We started this book, as we shall end it, by considering the risk of major and probably irreversible change in the earth's climate as a result of burning fossil fuels. Climatologists generally agree that through mechanisms which are complex, interactive, poorly understood, and possibly unpredictable even in principle, humankind's burning of fossil fuels (and some other major human activities) may already have perturbed global climate significantly (Hansen et al. 1981); if not, they may be about to do so before anything can be done to reduce the momentum; and at best, they will probably do so sometime in the first half of the next century unless world energy policies are markedly redirected over the next decade or two. The existence, seriousness, and probable irreversibility of the risk are hardly in question—only the timing and the exact effects. Chastening experience with other intricate interactive systems, notably ecosystems—where higher-order effects may dominate, lags are long, and an apparently small disturbance may suddenly shift behavior into a wholly different mode—suggests that if global climate does unravel, it may be in ways that we did not expect and are hard pressed to explain afterwards. What could happen, and what might trigger it, is not predictable with confidence and may never become so.

Prudence therefore requires forbearance in empirically testing the climatic limits of tolerance. If one's life may depend on the smooth functioning of a complex and mysterious machine, it is best not to stick monkeywrenches into it. By the time we understood its basic workings, if we ever did, it could be too late to stop in time to prevent serious harm. Major changes in the world energy system, albeit under necessity less stark than the oil pressures of today, have historically taken a half-century; but with the inexorable force of exponential energy growth which most studies still assume, major changes in climate may be set into motion in much shorter periods.

The stakes are illustrated by Holdren's (1980) calculation that if a CO_2-induced climatic change from burning about a tenth of the world's coal resource (increasing the concentration from about 338 to about 500 ppmv) risked a 10% chance of a famine killing 10^9 people, a 50% chance of a famine killing 10^8 people, and a 40% chance of no significant deaths, the risk would approach 7000 deaths/EJ. That is an order of magnitude more than the probable upper limit of air pollution deaths from burning that coal (3% sulfur, 80% release); it is about as big as the long-term hazard per primary EJ from uranium tailings' radon release (Ramsay & Russell 1978); and it is probably the largest energy-related risk to life other than nuclear war (Holdren 1980; Lovins & Lovins 1980).

We were, however, reluctant to estimate, unless we had to, the "value" of this climatic risk in order to offset it against the supposed extra cost of a "climatic insurance policy." (There is no theoretical basis or practical method for such a quantitative cost-risk-benefit comparison [Lovins 1977a; Junger 1976].) Fortunately, such an evaluation proved unnecessary, because we developed wholly separate arguments of economic and social efficiency which remain valid even if climatic risk is of no consequence. The policy with the lowest direct economic costs *also* happens to have the smallest climatic risks.

That may seem decisive, and recent trends toward a least-cost strategy are encouraging; but a rational outcome is not guaranteed, any more than the Swedish Parliament's 1980 adoption of a soft path means it will happen. Many persons influential in energy policy still maintain that *all* options must be pursued indiscriminately because we may need them someday—a holdover from the inflated demand forecasts made a decade ago. It is like saying that someone who wears both a belt and suspenders should buy a gross of safety pins and go about holding up his trousers. Yet it is a popular view among people who seek to avoid the harsh choices of cost-effectiveness.

It is also a dangerous view, because doing everything at once is not only unnecessary but impossible. Some options are incompatible with others. The cultural conditioning, accumulation of institutional barriers, and resource commitments of a high-energy, central-electrified future could easily push major soft-path implementation so far into the future that before we could get there, the transitional resources needed would already be irretrievably sunk. If we wish to use the cheapest remaining reserves of fossil fuels, and the relatively cheap money made from them, to capitalize the transition to a sustainable, climatically benign energy future, we have no time to lose. We need a *rapid* transition not only because the fossil fuel we shall have to use on the journey is becoming scarce and costly, but also because we dare not burn it for long. Our route must therefore be short, purposeful, and direct.

Even if climatic change is more remote than is now feared, we may have no second chance for the transition away from fossil fuel. Broad choices of energy strategy entail a certain irreversibility and mutual exclusivity (Lovins 1977; Nash 1979). Options compete for resources; ultimately they do not complement but devour each other. That competition is reflected in social structures and outlooks that are every bit as rigid as sunk costs. As these processes continue, reversals of direction may not become absolutely impossible, but at some point they do become prohibitively difficult. It is bad enough to envisage muddling along with an economically inefficient energy policy until the warning signs of CO_2-induced climatic change become unmistakable, then to try to procure the needed international agreement on non-fossil energy policies, reforestation, and so forth and hope it is not already too late. It would be far worse to discover, in such a case, that the flexibility of resources, institutions, and attitudes which briefly offered in the early 1980s an opportunity to pursue a low-climatic-risk energy strategy had been squandered on other, less successful, but extremely hard-to-reverse choices, and the low-risk option allowed to lapse. An offer of a good insurance policy against such a calamity—a policy, moreover, that repays its holder, rather than the reverse—is an offer too good to refuse.

It does not follow, however, that the insurance policy, even if procured, is an "all risks" policy. In fact, it is not at all comprehensive. Some energy-related climatic risks—CO_2 levels above about 370–400 ppmv, krypton-85 (Boeck et al. 1975), Beaufort Sea oil spills (Campbell & Martin 1973), perhaps Siberian river diversions—could be avoided, but many threats to climatic stability, notably deforestation, would remain (Bach et al. 1980). It is even possible that a badly managed biofuels program could exacerbate this problem: if not used as a vehicle for reforms to make farming and forestry sustainable (Jackson 1980; Lovins & Lovins 1981a), a biofuels program that only placed more stress on soil fertility could rapidly mine the enormous carbon inventory in the soil—the finely pulverized young coal in the belowground standing biomass and humus—and mobilize it into the air. Thus a climatic insurance policy against most energy risks still demands careful management, preventive medicine, and alertness to other threats to climatic health. It is not an excuse for relaxing vigilance, but rather a welcome relief from one set of climatic problems so that the talents of the world's climatologists can be focused on another set, related chiefly to our species' rapid erosion of the balance of the biosphere.

7.2. Conclusions

Our odyssey through the technicalities of an economically efficient energy future has now led us full circle to the questions raised by Table 1.1: what sort of trajectory of fossil-fuel burn is achievable, and how would the lowering of the trajectory from conventional projections reduce climatic risk?

We have argued that there is a technical, economic, and political possibility to stabilize and then steadily to reduce to approximately zero the global rate of burning fossil fuel—consistent with worldwide achievement of the most generously enthusiastic material goals—by substituting efficiency improvements and renewable sources which, unless grossly mismanaged, are climatically quite benign *per se* (Holdren et al. 1980). Such a policy would shrink the rate of burning fossil fuel further and faster than could be envisaged with any other policy (in the absence of catastrophic curtailments, e.g., by war, of oil supply or economic activity generally).

The time thus bought can be roughly estimated, subject to the simplifying assumptions stated, by referring once more to Table 1.1. The fifth line of the table, showing a constant rate of burning fossil fuel through 2025, corresponds to our *Gedankenexperiment* in which the entire globe, with 8 billion ultimate inhabitants, comes to look like the FRG in 1973, but using energy rather more efficiently, and substituting *no* additional renewable sources. In this case, the CO_2 level would attain 400 ppmv (a fifth above its present level) around 2025 and thereafter increase roughly linearly by about 1.4 ppmv/y.

But our analysis suggests that this is a worst case, for it takes no account of modern development policies or of presently available renewable sources that appear to be consistently cheaper and quicker to deploy than competing nonrenewable sources to replace the dwindling oil and gas. If, after an initial delay until 1985, the rate of burning fossil fuel were reduced by any combination of more rational development policies and more renewable supply, at a more or less linear rate of 2.2% simple interest per year, to 1 TW in 2025, then the integrated fossil-fuel burn during 1975–2025 would be reduced by 35% from its constant 8-TW level. If the rate of burn after 2025 were a constant 1 TW indefinitely, then the 400-ppmv CO_2 level would not be attained until about the year 2166. Nor is this result sensitive to our assumption that the fossil-fuel burn drops to 1 TW. Each TW by which the 2025 rate of fossil-fuel burn increased above 1 TW would add an integrated burn during 1985–2025 of only 20 TW-y, corresponding to an additional CO_2 increase by 2025 of only about 3.6 ppmv—less than the increment currently being added every three years!

In other words, what is important is the *fact* of a reduction in the rate of burning fossil fuel, not its exact pattern, which is immaterial for the purposes of this illustration. Given *any* combination of efficiency and renewables which steadily reduces the rate of burning fossil fuel to a few TW by 2025, then in the year 2025 the level of CO_2 in the atmosphere will be well below 400 ppmv, not about 450–550 as conventionally projected, and its rate of increase will be not 3.4–9.0 ppmv/y as conventionally projected but only about 0.2–0.5 ppmv/y. The *combination* of these two effects—a slower rate of increase from a lower base—will push the attainment of even a 400-ppmv level probably at least a century into the future. Reaching 450 ppmv at 1 TW would take over 460 years from now, rather than 30–50 years as conventionally forecast. Reaching 550 ppmv—nearly a doubling of the preindustrial level, corresponding to a global average warming of several C°—would take until the year 3000 AD, when one can safely assume the world will be very different in many ways more important than climate.

As a further example, to give more concreteness to the foregoing discussion, we combine into a single illustrative scenario our earlier calculations of a global demand scenario with strong economic growth, economically efficient energy use, and constant urban fraction (Tables 2.17 and 4.3) and the national renewable supply data discussed in Section 4.2.3. Because those data do not cover every country, we must make additional assumptions about renewables.[1] In combination, these assumptions yield a global renewable supply of 2.02 TW in 2000 and 4.27 TW in 2030—respectively 29% and 82% of the total primary

---

1. A reasonable set of assumptions is as follows:
- Energy demand per capita, and renewable supply as a fraction of primary demand, will be the same in Canada as in the United States (for which we take the average of the two SERI [1981] values for renewable supply). This undoubtedly underestimates the renewable supply from Canada (Lovins 1976a).
- Renewable supply as a fraction of primary demand will be the same in the EEC and elsewhere in Western Europe as the average for the UK, FRG, France and Denmark (which together represent 73% of the 1975 EEC energy demand). This is almost certainly conservative, as the outcome is heavily weighted by two countries—UK and FRG—whose renewable supply is considerably more difficult than is typical for Europe.
- Japanese renewables in 2030 will be the same as in 2010 (Tsuchiya 1980).
- Renewable supply as a fraction of primary demand will be the same in Australia, New Zealand, Israel, and South Africa as in Western Europe. This is again conservative in

demand of 7.07 TW and 5.23 TW (Table 2.17). On the further assumption of zero contribution from nuclear power or other sources besides fossil fuels and renewables—an assumption which maximizes CO_2 releases—fossil fuels would therefore be burned at a global rate of 5.05 TW in 2000 and 0.96 TW in 2030: essentially the hypothetical 1-TW case we have just discussed. If for simplicity we assume that all processes proceed at linear rates, then the average rate of change of the fossil-fuel burn is -0.107 TW/y from 1975 to 2000, -0.136 TW/y from 2000 to 2030, and (without Table 1.1's initial lag) -0.122 TW/y from 1975 to 2030.

To calculate the climatic impact of this burn trajectory, we need also to assume the future mix of fossil fuels, although our result is quite insensitive to the details of this assumption.[2] Given that plausible mix, the rate of increase of atmospheric CO_2 is about 0.8 ppmv/y in 2000 and 0.15 ppmv/y in 2030. The inte-

---

view of the enormous renewable potential of these countries, already expressed in programs in at least New Zealand and Israel.
- Renewable supply as a fraction of primary demand by the year 2000 will be the same in the average developing country as in the EEC. This is a reasonable and probably conservative representation of two competing forces—the developing countries' ability to buy the most cost-effective systems the first time, and the reluctance of many developed countries to promote or sell them. (For simplicity we omit here from both supply and demand estimates most of the present use of noncommercial fuel in developing countries. In fact, at least a tenth of global primary energy supply—and nearly the whole supply in some developing countries—comes today from renewable sources, mainly noncommercial fuels [Fritz 1981].)
- Renewable supply as a fraction of primary demand in Region II (USSR and Eastern Europe) will be the same in 2000 as in the EEC (viz., 22%)—less than the approximately 35% in Region I, but more likely in view of the East-bloc countries' climate and decision processes.
- Between 2000 and 2030, Region II and developing countries will achieve only 70% of the renewable supply fraction achieved in the EEC.
- No nation will produce more renewable energy supply than it needs for itself: that is, there will be no international trade in renewable energy, even to Regions II and IV–VII, which by 2030 are still 30% fossil-fueled. This is a major conservatism in view of our earlier discussion (Section 4.2.3) of national and regional surpluses.

2. The only reason for refining slightly the release coefficients shown in Table 1.2 is that at the low fossil-fuel demands assumed here, the rapid replacement of oil and gas by coal and synfuels which that Table assumes is no longer reasonable, and indeed there is no economic or logistical reason to build synfuel plants at all (Congressional Research Service 1981). For example, if the share of oil and gas in total fossil-fuel use remained constant at

grated fossil-fuel burn is 164 TW-y from 1975 to 2000 and 91 TW-y from 2000 to 2030, totalling 255 TW-y throughout the 55-year period: essentially the same as in the sixth case (linear decline to 1 TW) in Table 1.1.

Using our climatological assumptions from Section 1.2, this integrated burn implies a CO_2 level of about 371 ppmv in 2030, rising thereafter at a rate which decreases from 0.15 ppmv/y. A level of 400 ppmv is not attained until the year 2220 even if the "present technical limit" efficiencies are never reached and no further renewables are introduced after 2030: that is, if the fossil-fuel burn continues at a constant rate of 0.96 TW indefinitely after 2030. A modest improvement in either energy efficiency or the renewable contribution after 2030 would result in 400 ppmv *never* being attained. Furthermore, the integrated burns of oil and gas by 2030 are respectively about 73% and 44% of IIASA's estimates (Rogner & Sassin 1980) for resources *outside* Region VI, the Middle East and North Africa, at <$12/bbl (1975 $)—hardly a level of depletion that should be economically onerous. No contribution from nuclear fission or fusion is assumed at any time.

In short, IIASA [1981] outlined an energy policy which by 2030—despite massive reliance on coal and nuclear power—has largely depleted conventional hydrocarbons both in and outside the Middle East and at both low and high prices, and has reached CO_2 levels in the vicinity of 450 ppmv, rising by 50 ppmv every

---

their 1975 levels (Table 1.2), their burn in our 1975–2030 scenario would total respectively only 82% and 35% of the global resource *outside* Region VI (Middle East and North Africa) at prices <$12/bbl (1975 $), as estimated by IIASA (Rogner & Sassin 1980). This removes much of the present urgency to substitute. We nonetheless assume that the 1975 oil/gas/direct-coal shares shown in Table 1.2 (46%/24%/29%) will shift to 35%/30%/35% in 2000 and to 25%/35%/40% in 2030. This seems a plausible way of reflecting the political pressures to displace oil, the increasing dominance of metallurgical coal as a fraction of total coal use, and the importance of feedstocks. (Although our primary demand figures in this illustrative scenario exclude feedstocks, the feedstock demand may be as large as about 2 TW, and presumably this would offer a premium market for oil and gas.) These judgmental shares result in a CO_2 release coefficient which declines slightly—from 0.605 PgC/TW-y in 1975 to 0.602 in 2000 and to 0.601 in 2030. The effect is to reduce the release coefficient in 2030 by only 13% from that shown in Table 1.2—not a significant effect in view of the many other uncertainties in our simplified climatic model.

decade or so. In contrast, we have outlined a scenario which costs less, stretches oil and gas for centuries, dispenses with reliance on the Middle East and on nuclear power, and by 2030 has attained a CO_2 level barely 10% above today's level and rising by 5 ppmv every three decades or so. An economically efficient energy policy has thus made the CO_2 problem, and many others, virtually disappear.

As a final example of how least-cost energy investments can reduce climatic risks, we apply a more sophisticated climatic model to the four global energy scenarios compared in Table 7.1: the two IIASA scenarios (1981); the 16-TW Colombo & Bernadini (1979) scenario, assuming sources of energy supply proportional to those of the IIASA "low" scenario; and the efficiency scenario developed in Section 2.4 above, with supply as described in the four preceding paragraphs.

The "Index" row in Table 7.1 reveals that the IIASA scenarios indicate a 2.7- to 4.4-fold increase in global primary energy demand between 1975 and 2030, whereas our efficiency scenario, based on essentially identical projections of population and economic growth, shows a 27% reduction. The IIASA scenarios show a decrease in the fossil-fuel share of total energy supply, from 92% in 1975 to 67% or 69% in 2030, but the absolute rate of burning fossil fuel meanwhile increases from 7.6 TW to 14.9 or 24.8 TW. In contrast, our efficiency scenario, by relying on renewable sources that do not increase the amount of CO_2 in the air, reduces both the share of fossil fuel (from 92% to 18%) and the absolute rate of burning fossil fuel (from 7.6 TW to 0.96 TW). Further, by increasing the renewable share from 6% (1975) to 82% and dramatically improving end-use efficiency, the efficiency scenario dispenses with even the 1975 share of nuclear energy (2%), while the IIASA scenarios, increasing the renewable share only insignificantly to 10%, require an increase in the nuclear share from 2% to 23%.

Combining the data in Table 7.1 with reasonable assumptions[3] about the behavior of the earth's climatic system yields the

---

3. CO_2 emission data from 1860 to 1975 are calculated from UN statistics on coal, oil, and gas use plus gas flaring and cement use (Rotty, 1980), showing an average exponential growth rate of 3.4%/y. Atmospheric CO_2 concentrations are calculated from the

approximate patterns of CO_2 emission rate (in billion metric tons of carbon per year), CO_2 concentration (in parts per million by volume), and global average temperature change (in C°) shown in Figure 7.1a–d. Compared to the base year of 1975, the IIASA "high" scenario in 2030 shows a 3.4-fold increase in CO_2 emission rate and a 50% increase in CO_2 concentration, resulting in a 4-fold increase, from 0.4C° to 1.6 C°, above the 1860 tem-

---

emission data and the carbon cycle, starting with an assumed 1860 value of the order of 292 ppm and adjusted to measured values over the past twenty years (reaching 330 ppmv in 1975). The temperature changes shown for 1860–1970 are based on Mitchell's 1972 *Quaternary Research* compilation of actual measurements of surface air temperature averaged over latitudes 0–80°N.

The climatic projections from 1975 onwards first derive the atmospheric CO_2 concentration from the source terms in Table 7.1 and from the carbon-cycle models of Bacastow & Keeling (1973) and Zimen et al. (1978). Five CO_2 reservoirs are taken into account representing both sources and sinks, namely the atmosphere, the long- and short-lived biosphere, a mixing layer of the ocean, and the deep ocean. The fluxes between individual reservoirs are assumed proportional to the original carbon content. We neglect both deforestation and the boosting of biotic production by higher CO_2 levels, since Hampicke & Bach (1979) found these two effects were of similar magnitudes: in the global carbon budget, net atmospheric gains from tropical deforestation and the decomposition of soil organic matter are of order 1.3–4 GT/y, whereas CO_2-induced extra photosynthesis and temperate-zone reforestation may be of order 0.3–5 GT/y. Within this broad range of uncertainty, it is therefore not unreasonable to suppose that these two effects may roughly compensate.

A higher CO_2 concentration in the atmosphere will change the longwave radiation balance, increasing the surface and tropospheric air temperature to an extent currently best calculated through climatic models. Most models simply compare two assumed equilibria—an initial and a perturbed state. But in reality the changes in CO_2 concentration and temperature are not static but dynamic. We therefore use a time-dependent global ocean-land energy-balance climate model (Cess & Goldenberg, 1981) to estimate the global average surface air temperature as a function of time during 1975–2030. This horizontally averaged model uses a 70-m-deep oceanic mixing layer and a deep layer 3–4 km deep, exchanging heat by eddy diffusion. The global model incorporates the change in outgoing longwave radiation flux with surface temperature, the change in the solar radiation balance due to ice-albedo feedback, heat transport from land to ocean, and heat storage in the mixing layer. The calculated global temperature change is obtained by averaging sea and land air surface temperatures weighted by their areal shares. Comparison of climatic models with and without the effect of the ocean's heat capacity shows that the ocean could delay by about two decades (but not prevent) a global warming resulting from an increase in atmospheric CO_2.

The dates and temperatures calculated by this model are uncertain to about a factor of two. We present them, not to claim absolute precision, but rather as a qualitative comparison of the four energy scenarios' climatic consequences. That comparison is rather insensitive to the details of the climatic model used, even though the quantitative results are not.

Table 7.1 *Global rate of primary energy consumption (TW) and supply sources for a variety of energy scenarios, 1975–2030*

Source	Reference Year 1975		IIASA (1981) Scenarios						Colombo & Bernadini (1979) 16-TW (2030) Scenario[a]				Lovins et al. (1982) Efficiency Scenario			
			"High" Scenario				"Low" Scenario									
			2000		2030		2000		2030		2000		2030		2000		2030	
	TW	%	TW	%	TW	%	TW	%	TW	%	TW	%	TW	%	TW	%	TW	%
Oil	3.83		5.89		6.83		4.75		5.02		4.26		3.58		1.77		0.24	
Gas	1.51		3.11		5.97		2.53		3.47		2.27		2.48		1.51		0.34	
Coal	2.26		4.94		11.98		3.92		6.45		3.51		4.60		1.77		0.38	
Subtotal	7.60	92	13.94	83	24.78	69	11.20	82	14.94	67	10.04	82	10.66	67	5.05	71	0.96	18
Nuclear (LWR)	0.12		1.70		3.21		1.27		1.89		1.13		1.36		—		—	
Nuclear (breeder)	0		0.04		4.88		0.02		3.28		0.02		2.35		—		—	
Subtotal	0.12	2	1.74	10	8.09	23	1.29	10	5.17	23	1.15	10	3.71	23	0	—	0	0
Hydroelectricity	0.50		0.83		1.46		0.83		1.46		0.74		1.04		2.02			
Solar	0		0.10[b]		0.49[b]		0.09[b]		0.30[b]		0.08[b]		0.20[b]		}2.02		}4.27	
Other[c]	0		0.22		0.81		0.17		0.52		0.16		0.36					
Subtotal	0.50	6	1.15	7	2.76	8	1.09	8	2.28	10	0.98	8	1.60	10	2.02	29	4.27	82
Total	8.22		16.83		35.63		13.58		22.39		12.17		15.97		7.07		5.23	
Index	100		205		435		166		273		146		195		85		63	

a. Using the fuel shares of the "low" IIASA scenario. The 16-TW scenario is also known as the "zero growth scenario" because it assumes that the present average per-capita energy consumption will be the same in 2030 (despite extensive global development) but that a doubling in world population from 4 to 8 billion will lead to a doubling in total energy consumption.
b. Mostly soft solar (e.g. rooftop collectors) but also small amounts of centralized solar electricity.
c. Includes biogas, geothermal, commercial wood use.

Sources: Colombo and Bernadini (1979), Häfele et al. (1981), and this analysis.

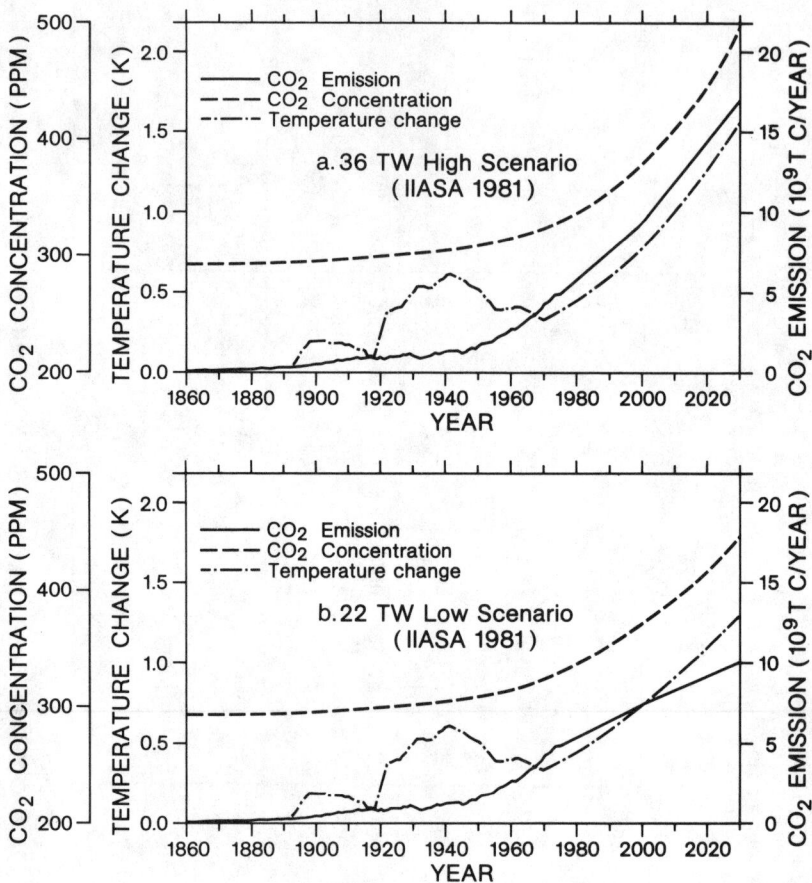

Figure 7.1 *Rate of CO_2 emission, CO_2 concentration, and global average temperature change ($C°$) for a variety of energy scenarios*

perature (Figure 7.1a). Over the same period, the efficiency scenario (Figure 7.1d) shows an 8-fold reduction in the CO_2 emission rate and only a 10% increase in CO_2 concentration, leading to a temperature increase of 0.8C°—only half as large. This increase is by no means insignificant, especially since the global average increase is about three or four times smaller than the increase in polar regions; but it is still a considerable savings, and by 2030 the rate of increase both of temperature and of the CO_2 concen-

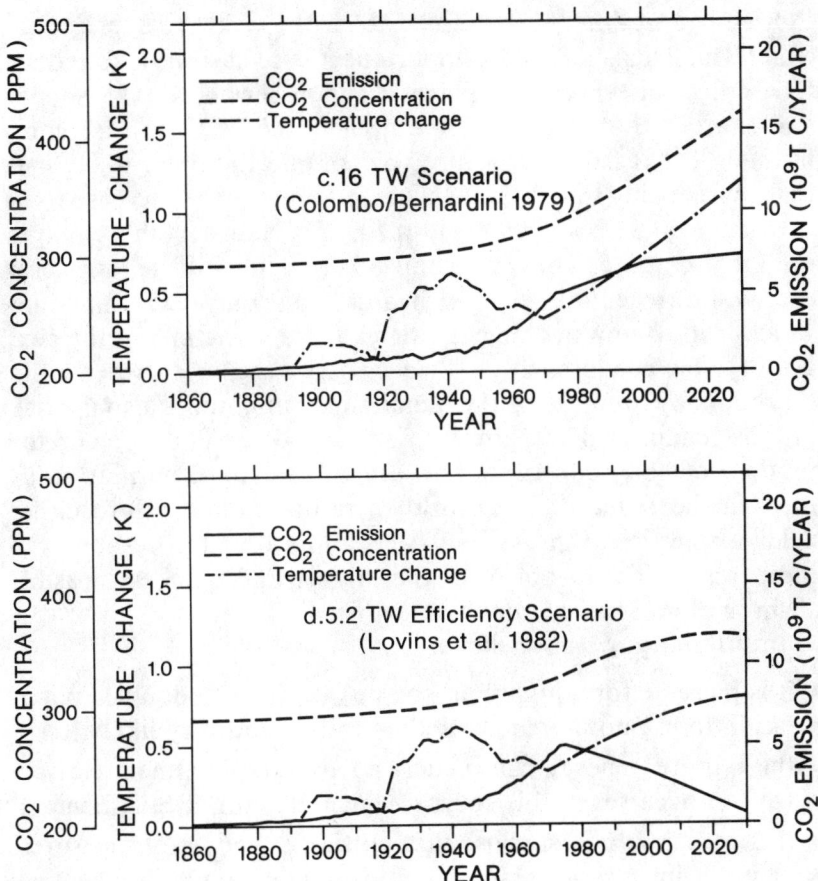

Figure 7.1 *(Continued) Rate of CO_2 emission, CO_2 concentration, and global average temperature change ($C°$) for a variety of energy scenarios*

tration which drives it is going down, not up.

Figure 7.1 also illustrates the importance of the inertia of the energy/climate system. The efficiency scenario assumes that efficiency improvements and renewables result in a steady decrease in the rate of CO_2 emission, starting in 1975, which is in fact nearly the case; yet it is not until about 2029, or some 54 years later, that the CO_2 concentration reacts with a slight decrease (Figure 7.1d), and it will take a very long time for the

CO_2 concentration then attained to drop back to its preindustrial level. This shows how important it is to start *now* to reduce CO_2 emissions through efficiency and renewables. The sooner and sharper that reduction, the greater the "CO_2 release commitment" that can be forever avoided, and the more likely it is that by reducing the ultimate average temperature increase to or even below $0.8C°$ we can avoid the risk of unacceptable changes in global climate. Thus, while all other visions of the long-term energy future entail short-term and continuing CO_2 emissions which lead to unwelcome climatic changes sometime in the next century, the economically efficient energy scenario presented in this book, by minimizing those emissions, minimizes also the risk of those changes. The degree of risk reduction will depend on the product of how *soon* times how *much* the rate of burning fossil fuel can be reduced. The world therefore cannot afford either delay or inefficient investment—the commitment of resources to measures which do not offer the largest and quickest possible saving in fossil fuel per dollar spent.

In summary, we have shown that

- it is possible for a highly prosperous world with doubled population to use no more energy than today, and very likely less;
- the required energy can be derived entirely or almost entirely from proven renewable sources which are climatically benign;
- it is reasonable to suppose that putting either set of measures, or both, into widespread global use might take a half-century or so starting now;
- either set of measures, or both in any combination, could rapidly reduce the rate of adding CO_2 to the atmosphere to such a low level that climatological concerns would probably become serious not in decades but in centuries, leaving ample leisure time to pursue fossil-fuel displacement further;
- this stretching of the time-scale of climatic risk is very insensitive to assumptions about the rates at which efficiency improvements or renewables are deployed, provided that their combined general effect is to reduce the rate of burning fossil fuel;
- provided this rate generally declines over the coming decades,

the bulk of the long-term contribution of CO_2 from burning fossil fuel will be made relatively soon, when policies are relatively foreseeable, so that any long-term delays in the more complete use of the fuel-displacing measures will be inconsequential;
- all these measures to displace fossil fuel make sense anyhow on other grounds and should be adopted even if climatic risks do not exist; and
- similar benefits cannot be achieved by other known means, notably by building coal or nuclear power stations or synfuel plants: on the contrary, such investments tend to slow down or even prevent the achievement of energy self-reliance, affordable and stable energy prices, and low climatic and other risks.

These conclusions tell the whole story. The key piece of information is that a pragmatic low-climatic-risk energy policy *is possible* and *is in any event preferable on other grounds, including economics*. Once citizens and their governments know that energy/climate risks are probably an artifact of an economically *in*efficient energy policy, they can decide whether to reject the risk or the inefficiency or both—and can then begin, knowing it is possible, to design a cheaper, safer future.

References

ALCOA CORPORATION 1973: *Chemical Engineering,* p. 61, 11 June.

ACEEE (American Council for an Energy-Efficient Economy) 1981: Proceedings of the August 1980 Santa Cruz Summer Study on Energy-Efficient Buildings, in press; available from J. Hollander, Lawrence Berkeley Laboratory, and summarized in SERI (1981).

ADLER, A. & MAYCOCK, P. 1981: briefings to the Energy Research Advisory Board, US Department of Energy, 5 February; transcript available from ERAB.

AG ENERGIEBILANZEN 1973: "Energiebilanzen für die Bundesrepublik Deutschland," Düsseldorf.

AYRES, R.U. 1974: "Substitution Possibilities and Problems in Regard to the Major Metals," TAD/RD/ENV/R.7, UNCTAD/UNEP, United Nations, Geneva.

AYRES, R.U. et al. 1974: *Critical Materials: A Problem Assessment,* IRT-348-R, International Research & Technology Corporation, 1501 Wilson Blvd., Arlington VA 22209.

AYRES, R.U. & NARKUS-KRAMER, M. 1976: "An Assessment of Methodologies for Estimating National Energy Efficiency," 76-WA/TS-4, American Society of Mechanical Engineers.

BACASTOW, R.B. & KEELING, C.D. 1973: "Atmospheric CO_2 and radiocarbon in the natural carbon cycle," at 86–134 in G.M. Woodwell and E.V. Pecan (eds.), *Carbon and the Biosphere,* US Atomic Energy Commission report CONF-720510, National Technical Information Service, Springfield VA 22161.

BACH, W. 1979: "Impact of Increasing Atmospheric CO_2 Concentrations on Global Climate: Potential Consequences and Corrective Measures," *Environment International* **2**:215–228, Pergamon, UK.

BACH, W. 1980: "Klimaeffekte anthropogener Energieumwandlung,"

Technische Mitteilungen **73**:496–510, June–July.

BACH, W. 1980a: "Climatic effects of increasing atmospheric CO_2 levels," *Experientia* **36**:796–806, Basel.

BACH, W. 1980b: "The CO_2 Issue—What Are the Realistic Options? An Editorial," *Climatic Change* **3**:3–5, Reidel, Dordrecht (Netherlands).

BACH, W. 1981: "Fossil Fuel Resources and Their Impacts on Environment and Climate," *International Journal of Hydrogen Energy* **6**:185–201, Pergamon, UK.

BACH, W. (coordinator) et al. 1980: "The carbon dioxide problem. An interdisciplinary survey," *Experientia* **36**:767–812 and 1017–1025, Basel.

BACH, W. & BREUER, G. 1980: "Wie dringend ist das CO_2-Problem?", *Umschau 80* **17**:520–525.

BACH, W., PANKRATH, J. & WILLIAMS, J. (eds.) 1980: *Interactions of Energy and Climate*, Reidel, Dordrecht (Netherlands).

BASILE, P.S. 1981: "Balancing Energy Supply and Demand: A Fifty-Year Global Perspective," *The Energy Journal* **2**(3):1–15, July, Tucson.

BAUERSCHMIDT, R. 1978: *Systemanalyse technologischer Einflussfaktoren*, Dissertation, Hannover.

BEN-DANIEL, D.J. & DAVID, E.E. JR. 1979: "Semiconductor Alternating Current Motor Drives and Energy Conservation," *Science* **206**:773f, 16 November.

BESANT, R.W., DUMONT, R.S. & SCHOENAU, G. 1978: "The Saskatchewan Conservation House: Some Preliminary Performance Results," Department of Mechanical Engineering, University of Saskatchewan, Saskatoon S1N 0W0, Canada.

BOECK, W.L. et al. 1975: *Bulletin of the American Meteorological Society* **56**:527f; see also *Science* **193**:195f (1976).

BOSQUET, P. 1978: report in special energy issue of *Que Choisir*, 7 rue Leonce-Reynaud, 75781 Paris 16, FFr. 10.

BOWRING, J. 1980: "Federal Subsidies to Nuclear Power: Reactor Design and the Fuel Cycle," March pre-publication draft, Financial & Industry Analysis Division, Office of Economic Analysis, Assistant Administrator for Applied Analysis, Energy Information Administration, US Department of Energy, Washington DC 20585; see also J. Omang, *Washington Post* A9, 8 April 1981.

BROOKS, D.B. 1981: *Zero Energy Growth for Canada*, McClelland & Stewart Ltd., Toronto.

BROOKS, D.B. 1981a: "Energy for Canada: The Soft Path," paper for Select Committee on Ontario Hydro Affairs, Legislative Assembly of Ontario; available from Dr. Brooks, 27–53 Queen St., Ottawa, Ont. K1P 5C5.

BROWN, N.L. & HOWE, J.W. 1978: *Science* **199**:651–657.

BRYENTON, R. 1980: July personal communication to M. Margolick (Department of Economics, University of British Columbia, Vancouver) regarding contents of p. 2, Appendix K, of "Passive Solar Remodelling—A Feasibility and Design Study to Assess Methods of Residential Passive Solar Energy

Collection and Utilization in Existing Housing," 1980 report to Ministry of Energy, Mines & Resources, Ottawa.
BUCHSBAUM, S. et al. 1979: *Jobs and Energy*, Council on Economic Priorities, New York.
Business Week 1981: "Energy Conservation: Spawning a billion-dollar business," cover story, pp. 58–69, 6 April.
BUTTI, K. & PERLIN, J. 1980: *A Golden Thread: 2500 Years of Solar Architecture and Technology*, Cheshire / Van Nostrand Reinhold, New York.
CAMPBELL, W.J. & MARTIN, S. 1973: "Oil and Ice in the Arctic Ocean: Possible Large-Scale Interactions," *Science* **181**:56–58.
CAPUTO, R. 1980: "Solar Energy for the Longer Term," ECE Seminar on Technologies Related to New Energy Sources, 8–12 December, UN Economic Commission for Europe, Geneva, and IIASA, Laxenburg. Expanded in "Solar Energy for the Next 5 Billion Years," IIASA Professional Paper PP-81-9, 1981.
CAPUTO, R. 1981: Presentation to Second International Conference on Soft Energy Paths, Rome, 16–18 January; proceedings to be published by IPSEP, 124 Spear St., San Francisco CA 94105.
CARLSON, R.C. et al. 1980: *California Energy Futures: Two Alternative Societal Scenarios and Their Energy Implications*, Research Report 31, February, prepared for California Energy Commission, 1111 Howe Ave., Sacramento CA 95825, by SRI International, 333 Ravenswood Ave., Menlo Park CA 94025.
CESS, R. & GOLDENBERG, S.D. 1981: "The effect of ocean heat capacity upon global warming due to increasing atmospheric CO_2," *Journal of Geophysical Research* **86**:498–502.
CGP (Commissariat du Plan) 1981: *Preparation du Huitième Plan 1981–1985: Rapport du Groupe de Travail. Prospective de la Consommation d'Énergie à Long Terme. Annexe: Secteur de l'Industrie*. La Documentation Française, Paris.
CHAPMAN, P.F. 1974: *Metals and Materials*, pp. 107–111, February.
CLAUSEN, C. & FRANKE, J. 1979: *Verstromungskosten von Brennstoffen in Leichtwasserreaktoren*, Bremen.
COLOMBO, U. & BERNADINI, O. 1979: "A Low Energy Growth 2030 Scenario and the Perspectives for Western Europe," report to Commission of the European Communities, Brussels.
COLOMBO, U. & BERNADINI, O. 1980: "A Low Energy Growth Scenario for the Year 2030," Comitato Nazionale per l'Energia Nucleare, Rome.
COMMITTEE ON GOVERNMENTAL AFFAIRS, US SENATE 1979: *Carbon Dioxide Accumulation in the Atmosphere, Synthetic Fuels and Energy Policy*, 30 July symposium, US Government Printing Office, Washington DC.
CONAES (Committee on Nuclear and Alternative Energy Systems, US National Research Council) Demand & Conservation Panel 1979: "US Energy Demand: Some Low Energy Futures," *Science* **200**:145–152, 18 April.

CONAES 1980: *Energy Choices in a Democratic Society*, Report of the Consumption, Location, and Occupational Patterns Group, Supporting Paper 7, National Academy of Sciences, Washington DC.

CONE, B.W. et al. 1980: *An Analysis of Federal Incentives Used to Stimulate Energy Production*, PNL-2410 Rev. II, Battelle Pacific Northwest Laboratory report to US Department of Energy, February.

CONGRESSIONAL RESEARCH SERVICE 1981: "Costs of Synthetic Fuels in Relation to Oil Prices," March report to USHR Committee on Science & Technology, Subcommittee on Energy Development & Applications, Serial B, USGPO, Washington DC.

CRAIG, P. et al. (eds.) 1978: *Distributed Energy Systems in California's Future: An Interim Report*, HCP/P7405-01/02, 2 vols., US Department of Energy, Washington DC.

DALY, H. 1978: *Steady-State Economics*, W.H. Freeman, San Francisco.

DARMSTADTER, J. (ed.) 1978: *International Comparisons of Energy Consumption*, Resources for the Future, Washington DC.

DITTERT, B., RUDOLPH, R. & HAAG, R. 1978: *Möglichkeiten der Energieeinsparung im Gebäudebestand*, Battelle Institut, Frankfurt/M.

DUMONT, R.S. 1981: personal communication, National Research Council of Canada, Division of Building Research, Saskatoon, Saskatchewan S7N 0W9.

DUMONT, R.S., BESANT, R.W. & KYLE, R. 1978: "Passive Solar Heating—Results from Two Saskatchewan Residences," Department of Mechanical Engineering, University of Saskatchewan, Saskatoon.

DUMONT, R.S., ORR, H.W., HEDLIN, C.P., & MAKOHON, J.T. 1980: "Measured Energy Consumption of a Group of Low-Energy Houses," paper to 1980 Annual Conference, Solar Energy Society of Canada, Vancouver, August.

DUNKERLY, J. et al. 1977: *How Industrial Societies Use Energy*, Resources for the Future / Johns Hopkins University Press, Baltimore, Maryland.

ECKHOLM, E. 1976: *Losing Ground: Environmental Stress and World Food Prospects*, W.W. Norton, New York.

EKETORP, S. 1980: "Future Steelworks," Royal Swedish Institute of Technology, Stockholm, September, 3 vols.

EKETORP, S., WIJK, O. & FUKAGAWA, S. 1981: "Direct Use of Coal for Production of Molten Iron," *Extraction Metallurgy*, in press.

ENCON CORPORATION 1978: descriptive literature on Hydro Place and Gulf Canada Square, Encon Corporation, 2200 Yonge St., Toronto, Ontario M4S 2O6.

ENQUÊTE-KOMMISSION 1980: *Zukünftige Kernenergiepolitik*, Bericht der Enquête-Kommission des Deutschen Bundestages, Bonn, Drucksache 8/4341.

FERCHAK, J.D. & PYE, E.K. 1981: "Utilization of Biomass in the U.S. for the Production of Ethanol Fuel as a Gasoline Replacement—I. Terrestrial Resource Potential," *Solar Energy* **26**:9–16.

FERCHAK, J.D. & PYE, E.K. 1981a: "Utilization of Biomass in the U.S. for the Production of Ethanol Fuel as a Gasoline Replacement—II. Energy Requirements, with Emphasis on Lignocellulosic Conversion," *Solar Energy* **26**:17–25.

FfE (Forschungsstelle für Energiewirtschaft) 1974: "Ermittlung des kumulierten spezifischen Energieaufwandes zur Herstellung von Weissblechbehältern," EG-Studienauftrag Nr. 145-74-ECIC, München.

FfE 1977: "Technologien zur Einsparung von Energie im Endverbrauchssektor Verkehr," Mai, München.

FfE & RWE 1977: "Zentrale Wärmerückgewinnung aus dem Warmwasserverbrauch in Mehrfamilienhäusern," BMFT Forschungsvorhaben ET 5012, bearbeitet durch FfE, München, & RWE Anwendungstechnik, Essen.

FICHTNER BERATENDE INGENIEURE (Stuttgart) 1977: *Technologien zur Einsparung von Energie*, Bundesministerium für Forschung und Technologie, Bonn.

FISHELSON, G. & LONG, T.V. II 1977: "An International Comparison of Energy and Materials Use in the Iron and Steel Industry," Committee on Public Policy Studies, University of Chicago.

FOLEY, G. 1979: "Economic Growth and the Energy Future," report ENV/SEM.11/R.21 to 3–8 December Ljubljana meeting of Senior Advisors to ECE Governments on Environmental Problems, UNEP/ECE Regional Seminar on Alternative Patterns of Development and Lifestyles, UN Economic Commission for Europe, Geneva.

FRITZ, M. 1981: *Future Energy Consumption of the Third World*, Pergamon.

GILES, W.L. 1979: "Expanded Applications Wide Base Radial Truck Tire," Society of Automotive Engineers Paper #791044.

GIRY 1978: report available from Les Amis de la Terre, 72 rue du Chateau d'Eau, 75010 Paris.

GLEASON, J. 1981: "District Heating: Choice between 'Hard' and 'Soft' Path Technology," *Self-Reliance* **25**:1,6–7, January–February, Institute for Local Self-Reliance, 1717 18th St. NW, Washington DC 20009.

GLIDDEN, W.T. JR. & HIGH, C.J. 1980: *The New England Energy Atlas*, Resource Policy Center, Thayer School of Engineering, Dartmouth College, Hanover NH 03755.

GOELLER, H. E., & WEINBERG, A. M., 1975: "The Age of Substitutability," Institute for Energy Analysis (Oak Ridge) and Oak Ridge National Laboratory, 18 September, Oak Ridge, Tennessee.

GOLDEMBERG, J. 1978: "Energy Strategies for Developing and Less Developed Countries," PU/CES 70, Center for Energy & Environmental Studies, Princeton University, Princeton NJ 08540.

GORMAN, R. & HEITNER, K. L. 1980: "A Comparison of Costs for Automobile Energy Conservation *vs.* Synthetic Fuel Production," TRW Energy Systems Group, 8301 Greensboro Drive, McLean VA 22102.

GRAY, C. & VON HIPPEL, F. 1981: "The Fuel Economy of Light Vehicles," *Scientific American*, pp. 48-59, May.

GRENON, M. 1975: at 42 and 4 (F3) in W. Häfele et al., *Second Status Report on the IIASA Project on Energy Systems 1975*, IIASA, Laxenburg, Austria.
GRETHLEIN, H. E., CONVERSE, A. O., MCPARLAND, J. J., & SMITH, P. C. 1980: "Acid Hydrolysis of Cellulosic Biomass in a Continuous Plug Flow Reactor," final report to USDOE/SERI, Contract #EG 77-S-01-4061, Thayer School of Engineering, Dartmouth College, Hanover NH 03755, October.
GROUPE DE BELLEVUE 1978: "Project Alter" report, 85 boul. de port Royal, 75031 Paris, FFr. 8.
GYFTOPOULOS, E. P., LAZARIDIS, L. J. & WIDMER, T. F. 1974: *Potential Fuel Effectiveness in Industry*, Ballinger for The Energy Policy Project, Cambridge, Massachusetts.
HÄFELE, W. 1979: "Langfristige Strategien zur Energieversorgung," 19 June paper to ACHEMA meeting, Frankfurt/M, IIASA, Laxenburg, Austria.
HAKE, B. 1980: *Ölkrisenprogramm für Hausbesitzer*, Wiesbaden.
HAMPICKE, U. & BACH, W. 1979: "Die Rolle terrestrischer Ökosysteme im globalen Kohlenstoff-Kreislauf," report #104 02 513 to the German Federal Environmental Agency (Berlin), Essen/Münster.
HANN, W. 1979: *Frankfurter Allgemeine Zeitung*, den 8. März, Nr. 178, S. 11.
HANNON, B. 1976: "Energy and Labor Demand in the Conserver Society," Center for Advanced Computation, University of Illinois at Urbana-Champaign.
HANSEN, J., JOHNSON, D., LACIS, A., LEBEDEFF, S., LEE, P., RIND, D. & RUSSELL, G. 1981: "Climate Impact of Increasing Atmospheric CO_2," NASA Institute for Space Studies, Goddard Space Flight Center, New York NY 10025.
HAYES, D. 1977: "Energy for Development: Third World Options," paper 15, Worldwatch Institute, 1776 Massachusetts Ave., Washington DC 20036.
HEIN, K. 1979: *Blockheizkraftwerke*, C.F. Müller Verlag, Karlsruhe.
HÖGLUND, I., JOHNSSON, B., & LAGERSTRÖM, J. 1981: *Ulvsundaprojektet: Effektivare energianvändning i äldre byggnader: Ettap I*, Rap. 1981:T5, Byggforskningsrådet, Stockholm, ISBN 91-540-3411-6.
HOLDREN, J. P. 1980: "Cross-Cutting Issues in Integrated Environmental Assessment of Energy Alternatives: Distribution of Costs and Benefits," ERG-WP-80-10, March, Energy & Resources Group, 100 T-4, University of California, Berkeley CA 94720.
HOLDREN, J.P. et al. 1980: "Environmental Aspects of Renewable Energy Sources," ERG-80-1, loc.cit.
HOLLANDS, K.G.T., & ORGILL, J.F. 1977: *Potential for Solar Heating in Canada*, Report 77-01 to Division of Building Research, National Research Council of Canada, Project 4107-1,2, University of Waterloo Research Institute, Waterloo, Ontario.
HOWE, J.W. 1977: *Energy for the Villages of Africa*, Overseas Development Council, Washington DC.
IEA (International Energy Agency) 1978: *Workshop on Energy Data of Developing Countries*, December, OECD, Paris.

IIASA (International Institute for Applied Systems Analysis) 1978: *Energy Systems Status Report,* IIASA, 2361 Laxenburg, Austria.

IIASA 1981: *Energy in a Finite World: A Global Systems Analysis,* 2 vols., February, Energy Systems Group, IIASA, Laxenburg, Austria, and Ballinger, Cambridge MA 02138.

Institute of Energy Economics 1979: *Present State and Future Potential of Energy Conservation in Japan,* IEE, No. 10 Mori Bldg., 1-18-1 Toranomon, Minato-ku, Tokyo, $15 English summary, 33 pp., $65 Japanese text, 320 pp. Summarized in *Soft Energy Notes* 3(1):12–13, February 1980.

InterTechnology 1977: *Analysis of the Economic Potential of Solar Thermal Energy to Provide Industrial Process Heat,* 3 vols., COO-2829, February report to ERDA Division of Solar Energy, InterTechnology Corporation, 100 Main St., Warrenton VA 22186; available from National Technical Information Service, Springfield VA 22161.

IPSEP (International Project for Soft Energy Paths) 1980: Proceedings of the First International Conference on Soft Energy Paths, Rome, May 1979; available from IPSEP, 124 Spear St., San Francisco CA 94105.

IPSEP 1981: Proceedings of the Second International Conference on Soft Energy Paths, Rome, January 1981, in preparation; available from IPSEP.

IWW (Institut für Weltwirtschaft) 1977: "Der internationale Strukturwandel," IWW, Kiel.

JACKSON, W. 1980: *New Roots for Agriculture,* Friends of the Earth, 124 Spear St., San Francisco CA 94105.

JACKSON, W. & BENDER, M. 1980: "New Roots for American Agriculture," report to OTA, The Land Institute, Rt. 3, Salina KS 67401.

JACKSON, W. & BENDER, M. 1980a: "American Food: S/oil and Water" and "Saving Energy and Soil," *Soft Energy Notes* 3(6):3–7, December 1980/January 1981.

JOHANSSON, T.B. & STEEN, P. 1978: *Solar Sweden,* Secretariat for Future Studies, Fack, 10310 Stockholm.

JUNGER, P. 1976: "A Recipe for Bad Water: Welfare Economics and Nuisance Law Mixed Well," *27 Case Western Reserve Law Review* 3–335, Fall.

KAHN, E. 1979: "The Compatibility of Wind and Solar Technology With Conventional Energy Systems," *Annual Review of Energy* 4:313–352, Annual Reviews, Inc., Palo Alto CA 94306.

KELLOGG, H.H. 1977: "Conservation and Metallurgical Process Design," 13th Wernher Memorial Lecture, Institute of Mining & Metallurgy, London.

KELLOGG, W.W. & SCHWARE, R. 1981: *Climate Change and Society,* Aspen Institute for Humanistic Studies/Westview Press, Boulder CO 80301.

KEYFITZ, N. 1977: "Population of the World and its Regions, 1975 to 2030," WP-77-7, International Institute for Applied Systems Analysis, Laxenburg, Austria.

KIHLSTEDT, P.G. 1975: *Scandinavian Journal of Metallurgy* 4:145–149.

KIHLSTEDT, P.G. 1977: "Samhällets Råvaruförsörjning under Energibrist," IVA-Rapport 112, Ingenjörsvetenskapsakademien, Stockholm.

KOMANOFF, C. 1981: *Power Plant Cost Escalation*, March, Komanoff Energy Associates, 333 West End Ave., 14th floor, New York NY 10023.

KRAUSE, F. 1980: *Wirtschaftswachstum bei sinkendem Energieverbrauch—Szenarien für die Energieversorgung der Bundesrepublik ohne Erdöl und Uran*, ÖKO-Institut, Freiburg.

KRAUSE, F. 1981: "The Industrial Economy: An Energy Barrel Without a Bottom?", paper to Second International Conference on Soft Energy Paths, Rome, 16–18 January, available from IPSEP, 124 Spear St., San Francisco CA 94105.

KRAUSE, F., BOSSEL, H., & MÜLLER-REISSMAN, K.F. 1980: *Energiewende*, Fischer, Frankfurt/M.

LADISCH, M.R. et al. 1978: "Cellulose to sugars: new path gives quantitative yield," *Science* **201**:743–745.

LEACH, G. 1979: "Report of the Energy Resources Working Group," international conference *Agricultural Production: R&D Strategies for the 1980s*, Bonn, October; available from IIED (below).

LEACH, G. et al. 1979: *A Low Energy Strategy for the United Kingdom*, International Institute for Environment & Development, 10 Percy St., London W1P 0DR, England, and Science Reviews Ltd., London.

LEDERGERBER, E. 1979: *Wege aus der Energiefalle: Handlungsspielräume und Strategien für eine unabhängigere Energieversorgung der Schweiz*, Verlag Rüegger, 8253 Diessenhofen, Schweiz.

LEGER, E.H. & DUTT, E.S. 1979: "An Affordable Solar House," 4th National Passive Solar Conference, Kansas City; available from Vista Homes, PO Box 95, E. Pepperell MA 01437.

Les Amis de la Terre 1978: "Tout Solaire," 72 rue du Chateau d'Eau, 75010 Paris.

LEWIS, C. 1980: *Energy World*, pp. 2–8, January, Institute of Energy, London.

LINDSTRÖM, O. 1979: "Bio-fuels make Sweden independent of oil and nuclear energy," October, Dept. of Chemical Technology, Royal Institute of Technology, 10044 Stockholm.

LOVINS, A.B. 1973: *Openpit Mining*, Earth Island Ltd., London; available from FOE, 124 Spear St., San Francisco CA 94105.

LOVINS, A.B. 1976: "Long-Term Constraints on Human Activity," *Environmental Conservation* **3**(1):3–14, Geneva.

LOVINS, A.B. 1976a: "Exploring Energy-Efficient Futures for Canada," *Conserver Society Notes* **1**(4):5–16, May/June, Science Council of Canada, Ottawa.

LOVINS, A.B. 1977: *Soft Energy Paths*, Ballinger, Cambridge, Massachusetts, and Harper & Row Colophon, New York; also published in German as *Sanfte Energie*, Rowohlt, Reinbek bei Hamburg, 1978.

LOVINS, A.B. 1977a: "Cost-Risk-Benefit Assessments in Energy Policy," **45** *George Washington Law Review* 911–943, August.

LOVINS, A.B. 1978: "Re-Examining the Nature of the ECE Energy Problem," ECE (XXXIII)/2/I.G., UN Economic Commission for Europe, Geneva,

January; reprinted in *Energy Policy* 7:178-198, September 1979.
LOVINS, A.B. 1978a: Third Regents' Lecture, University of California at Berkeley, 25 May; unedited transcript available with Vu-Graphs from IPSEP, 124 Spear St., San Francisco CA 94105.
LOVINS, A.B. 1978b: "Soft Energy Technologies," *Annual Review of Energy* 3:477-517, Annual Reviews, Inc., Palo Alto CA 94306.
LOVINS, A.B. 1978c: "Lovins on Energy Costs," *Science* 201:1077-1078.
LOVINS, A.B. 1978d: "How To Finance the Energy Transition," *Not Man Apart*, pp. 8-10, mid-Sept./Oct., Friends of the Earth, San Francisco.
LOVINS, A.B. 1979: "Energy: Bechtel Cost Data," *Science* 204:124-129.
LOVINS, A.B. 1980: "Economically Efficient Energy Futures," at pp. 1-31 in (Bach, Pankrath, & Williams 1980).
LOVINS, A.B. 1980a: Preface to Japanese edition of *Soft Energy Paths*, Jiji Tsushinsha, Tokyo.
LOVINS, A.B. 1981: "Expansio ad Absurdum," *The Energy Journal* 2(3):25-34, October, Tucson.
LOVINS, A.B. 1981a: Supplementary information supplied for the record, 3 April 1981 hearings, *Energy Production vs. Energy Efficiency in the Utility Industry*, Subcommittee on Energy Conservation & Power, Committee on Energy & Commerce, US House of Representatives, Washington DC.
LOVINS, A.B. 1981b: "Electric Utility Investments: *Excelsior* or Confetti?", *Journal of Business Administration* 12(2):91-114, Vancouver.
LOVINS. A.B. 1981c: "How to Keep Electric Utilities Solvent," memorandum to Secretary of the Treasury, 26 February, in press, *The Energy Journal*.
LOVINS, A.B. 1981d: Surrebuttal testimony of 14 August in Limerick Investigation, Docket #I-80100341, Pennsylvania Public Utility Commission, Harrisburg.
LOVINS, A.B. & LOVINS, L.H. 1980: *Energy/War: Breaking the Nuclear Link*, Friends of the Earth, San Francisco, and Harper & Row, New York (1981); *Atomenergie und Kriegsgefahr*, Rowohlt, Reinbek bei Hamburg 1981.
LOVINS, A.B. & LOVINS, L.H. 1981: *Energy Policies for Resilience and National Security*, report to President's Council on Environmental Quality and Federal Emergency Management Agency, May, contract DCPA 01-79-C-0317, FEMA Work Unit 4351C, November; revised edition, *Brittle Power: Energy Strategy for National Security*, in press, Brick House, spring 1982.
LOVINS, L.H. & LOVINS, A.B. 1981a: "Biomass Fuels: Options and Obstacles," typescript available from IPSEP, 124 Spear St., San Francisco CA 94105.
MAKHIJANI, A. 1976: "Energy Policy for the Rural Third World," International Institute for Environment & Development, 10 Percy St., London W1P 0DR.
MAKHIJANI, A. & POOLE, A. 1975: *Energy and Agriculture in the Third World*, Ballinger for The Energy Policy Project, Cambridge MA 02138.
MALTE (Miljörörelsens Alternative Energiplan) 1977: *Huvudrapport och Bilagorna*, 3 vols., Box 24023, 40022 Göteborg, Sweden.

MARGEN, P. 1980: "Economics of Solar District Heating," *Sunworld* **4**(4):128-134, International Solar Energy Society, Pergamon, UK.
MARSHALL, E. 1980: "Energy Forecasts: Sinking to New Lows," *Science* **208**: 1353-1356, 20 June.
MAYCOCK, P.D. & STIREWALT, E.N. 1981: *Photovoltaics,* Brick House, Andover MA 01810.
MEADOWS, D. 1981: "A Critique of the IIASA Energy Models," *The Energy Journal* **2**(3):17-28, July, Tucson.
MEYER, N., NØRGÅRD, J.S., BLEGÅ, S. et al. 1977-79: Reports of the DEMO-Project, Physics Lab. III, Technical University of Denmark, 2800 Lyngby.
MEYER-ABICH, K. (Hrsg.) 1978: "Wirtschaftspolitische Steuerungsmöglichkeiten zur Einsparung von Energie durch alternative Technologien," Arbeitsgruppe Umwelt Gesellschaft Energie, Universität Essen.
Mitsubishi Research Institute, Inc.: "Assessing Incentives Created by the Japanese Government to Stimulate Energy Production," final report to US Department of Energy, ca. 1978, MRI Inc., 1-8-1 Yuraku-cho, Chiyoda-ku, Tokyo 100.
MURGATROYD, W. & WILKINS, C.B. 1976: "The Efficiency of Electric Motive Power in Industry," *Energy* **1**:337-346.
NAS (National Academy of Sciences/National Research Council) 1969: *Resources and Man*, W.H. Freeman, San Francisco.
NASH, H. (ed.) 1979: *The Energy Controversy: Soft Path Questions and Answers*, Friends of the Earth, 124 Spear St., San Francisco CA 94105.
NEUMANN, F., LEU, H., PTACH, R. & BRUSA, U. 1975: "Das BBC-Brusa-Verfahren zum Schmelzen von Stahl," *Stahl und Eisen* **95**:1ff.
New Shelter 1980: feature on passive solar cooling, July/August, Rodale Press, Emmaus, Pennsylvania.
NIEHAUS, F. 1980: "The impact of different energy options on atmospheric CO_2 levels," at 792-796 in Bach et al. (1980).
NØRGÅRD, J.S. 1979: "Improved Efficiency in Domestic Appliances," *Energy Policy*, March, pp. 43-56.
NØRGÅRD, J.S. 1979a: *Husholdninger og Energi*, Polyteknisk Forlaget, København.
NØRGÅRD, J.S. 1981: Address of 15 January, in *Soft Energy Notes* **4**(1):2-3 (see also :5), February/March.
NORMAN, C. 1981: "Energy Conservation: The Debate Begins," *Science* **212**:424-426, 24 April.
OFFICE OF MANAGEMENT & BUDGET 1975: *Interagency Report on U.S. Government Export Promotion Policies and Programs*, rev. ed., April, Washington DC.
OLIVIER, D. et al. 1981: *Low Energy Scenario for Great Britain, I and II*, prepared for Energy Technology Support Unit, Harwell, UK Atomic Energy Authority, by Earth Resources Research Ltd., 258 Pentonville Rd., London N.1, England, to be published.
ORTH, G. 1976: "Untersuchungen zu den wirtschaftlichen Möglichkeiten einer Verringerung der Energieintensität bei elektrischen Haushaltsgeräten," Ber.

ET-5003A, Elektro-Wärme-Institut, e.V., Essen.
OTA (Office of Technology Assessment, US Congress) 1977: *Application of Solar Technology to Today's Energy Needs*, 2 vols.
OTA 1979: *Materials and Energy from Municipal Waste*, OTA-M-93, July, Washington DC 20510.
OTA 1979a: *Technical Options for Conservation of Metals*, September, US Government Printing Office #052-003-00705-3, September.
OTA 1980: *Energy From Biological Processes*, 2 vols., OTA-E-124, July.
PALMITER, L. & MILLER, B. 1979: "Report on the October 13–14, 1979 Saskatchewan Conference on Low Energy Passive Housing," National Center for Appropriate Technology, Butte MT 59701.
PATTERSON, D. 1980: "Fuel Alcohol Production in the United States: A Status Review and Evaluation," in Joint Economic Committee, US Congress, *Farm and Forest Produced Alcohol: The Key to Liquid Fuel Independence*, US Government Printing Office, Washington DC.
PETER, R.W. 1977: *Ein Nationaler Energiesparplan*, Gottlieb Duttweiler-Institut, Rüschlikon-Zürich.
POMERANCE, D., KOSTER, F., WAGSHAL, P. et al. 1979: *Franklin County Energy Study: A Renewable Energy Scenario for the Future*, 600 pp., $12, Franklin County Energy Project, Box 548, Greenfield MA 01301.
President's Council on Environmental Quality 1981: *Global Energy Futures and the Carbon Dioxide Problem*, January, US Government Printing Office, Washington DC.
President's Council on Environmental Quality and Department of State 1980: *The Global 2000 Report to the President: Entering the Twenty-First Century*, 2 vols., July, US Government Printing Office, Washington DC.
President's Council on Environmental Quality and Department of State 1981: *Global Future: Time to Act: Report to the President on Global Resources, Environment and Population*, January.
RAETZ, K.-H. 1979: "Konvex-Sonnenlichtkollektor," *PTB-Mitteilungen* **89**:217–218, März, Physikalisch-Technische Bundesanstalt, Braunschweig.
RAETZ, K.-H. 1979a: "Der Konvexkollektor," *TAB* IX. 79; available from author, Gassnerstr. 12, 3300 Braunschweig.
RAINS, R.K. & KADLEC, R.H. 1970: "The Reduction of Al_2O_3 to Aluminum in a Plasma," *Metallurgical Transactions* **1**(6):150lff.
RAMSAY, W. & RUSSELL, M. 1978: *Public Policy* **26**:387–403, Harvard University.
REDDY, A.K.N. 1978: *Bulletin of the Atomic Scientists*, May, pp. 28–33, and December, pp. 54–55.
REDDY, A.K.N. 1980: "Alternative Energy Policies for Developing Countries: A Case Study of India," at 289-351 in R.A. Bohm et al., eds., *World Energy Production and Productivity*, Ballinger, Cambridge MA 02138; updated MS from IPSEP or from author at Indian Institute of Science, 560012 Bangalore.
REDDY, A.K.N. & PRASAD, K.K. 1977: "Technological Alternatives and the Indian Energy Crisis," *Economic and Political Weekly* **12**

(33-34SN):1465–1602, August, Bombay.

REENTS, H. 1977: "Die Entwicklung des sektoralen End- und Nutzenergiebedarfs in der Bundesrepublik Deutschland," Ber. 1452, KFA Jülich GmbH, August.

RODBERG, L. 1978: *The Employment Impact of the Solar Transition,* Joint Economic Committee, US Congress, US Government Printing Office #052-04915-4, Washington DC.

ROGNER, H.-H. & SASSIN, W. 1980: "High Energy Demand and Supply Scenario," at pp. 33–52 in Bach, Pankrath, & Williams (1980).

ROMIG, F. & LEACH, G. 1977: *Energy Conservation in UK Dwellings: Domestic Sector Survey and Insulation,* International Institute for Environment & Development, 10 Percy St., London W1P 0DR, England.

ROSENFELD, A.H. et al. 1980: "Building Energy Use Compilation and Analysis (BECA): An International Comparison and Critical Review," LBL-8912, Lawrence Berkeley Laboratory, Berkeley CA 94720.

ROSS, M.H. & WILLIAMS, R.H. 1981: *Our Energy: Regaining Control,* McGraw-Hill, New York.

ROTTY, R.M. 1979: presentation at 185–194 in (Committee on Governmental Affairs 1979).

ROTTY, R.M. 1980: "Past and future emission of CO_2," *Experientia* **36**(7):781–783, Basel.

RTM (Resource Technology Management Corp.) 1981: *Alternative Energy Data Summary™ for the United States, 1975–1980.* Vol. 1. RTM, 714A S. 15th St., Arlington VA 22202.

RUSSELL, M.C. 1981: "A Summary of Residential Photovoltaic System Economics," 6 January, MIT Lincoln Laboratory, Lexington MA 02173.

SAE (Society of Automotive Engineers) 1980: "Radial Ply Truck Tires Increase Fuel Economy," *Automotive Engineering,* p. 69, January.

SANT, R. 1979: "The Least-Cost Energy Strategy: Minimizing Consumer Costs Through Competition," Energy Productivity Center, Mellon Institute, Suite 1200, 1925 N. Lynn St., Arlington VA 22209; summarized in *Harvard Business Review* 6ff, May–June 1980.

SANT, R. et al. 1981: "Eight Great Energy Myths: The Least-Cost Energy Strategy 1978–2000," Energy Productivity Center, loc. cit.

SCHACHTER, M. 1979: *The Job Creation Potential of Solar and Conservation: A Critical Analysis,* Office of Policy Evaluation, US Department of Energy.

SCHÄFER, H. 1971: "Analyse des Kraftbedarfs und seiner Nutzungsgrade in der Bundesrepublik," *Brennstoff-Wärme-Kraft* **23**(5).

SCHIPPER, L. 1978: "Energy Use and Conservation in Industrialized Countries," LBL-7872, Lawrence Berkeley Laboratory, Berkeley CA 94720.

SCHIPPER, L. & KETOFF, A. 1980: "International Residential Energy End Use Data: Analysis of Historical and Present Day Structure and Dynamics," LBL-10587, September, Lawrence Berkeley Laboratory, Berkeley CA 94720, as updated (pers. comm.) January 1981.

SCHIPPER, L. & LICHTENBERG, A.J. 1976: "Efficient Energy Use and Well-

Being: The Swedish Example," *Science* **194**:1001–1013, 3 December.

SCHIPPER, P.G. & TUININGA, E.-J. 1979: Annex 6 to J. Saint-Geours et al., "In Favour of an Energy-Efficient Society," EEC, Brussels.

SCHMITZ, K. & VOSS, A. 1980: "Aktuelle Beiträge zur Energiediskussion Nr. 2," Bericht Jül.-spez.-73, April, KFA Jülich GmbH.

SCHNEIDER, S. H. 1980: "The CO_2 Problem: Are There Policy Implications—Yet?", Editorial, *Climatic Change* **2**:203–205.

SCHNEIDER, S. H. 1981: Carbon Dioxide and Climate: Research on Potential Environmental and Societal Impacts," 31 July testimony to Subcommittee on Natural Resources, Agricultural Research and Environment, Committee on Science and Technology, US House of Representatives.

SCHNEIDER, S. H. and CHEN, R.S. 1981: "Carbon Dioxide Warming and Coastline Flooding: Physical Factors and Climatic Impact," *Annual Review of Energy* **5**:107–140, Annual Reviews, Inc., Palo Alto CA 94306.

SCHÜRLE, N., KLAISS, & SCHULZ, K.H. 1977: "Wirtschaftlichkeitsvergleich Zukünftiger Raumheizungsysteme für den Sektor Private Haushalte," Ber. IKE K-54-6, Institut für Kernenergetik, Stuttgart.

SCHWEIZERISCHE ENERGIESTIFTUNG 1978: *Jenseits der Sachzwänge*, SES, Auf der Mauer 6, 8001 Zürich.

SCIENCE COUNCIL OF CANADA 1977: *Canada as a Conserver Society*, Ottawa.

SEIFFERT, U. et al. 1979: "Improvements in Automotive Fuel Economy," paper to First International Automotive Fuel Economy Conference, November, Washington DC.

SEIFFERT, U. & WALZER, P. 1980: "Development Trends for Future Passenger Cars," paper to 5th Automotive News World Congress, August preprint, from Research Division, Volkswagenwerk AG, Wolfsburg.

SERI (Solar Energy Research Institute) 1980: *Community Energy Self-Reliance*, Proceedings of the First Conference on Community Renewable Energy Systems, Boulder, Colorado, 20–21 August 1979, SERI/CP-354-421, July, available from SERI, 1617 Cole Blvd., Golden CO 80401; the proceedings of the analogous August 1980 conference, in Seattle, are in press.

SERI 1981: *Building a Sustainable Energy Future*, draft, preprinted (#97-K) by Committee on Energy & Commerce, U.S. House of Representatives, 2 vols., April, US Government Printing Office, Washington DC; reprinted by Brick House Publishing Co., 34 Essex St., Andover MA 01810.

SHACKSON, R.H. & LEACH, H.J. 1980: "Using Fuel Economy and Synthetic Fuels to Compete with OPEC Oil," Energy Productivity Center, Mellon Institute, Suite 1200, 1925 N. Lynn St., Arlington VA 22209, 18 August.

SHICK, W. 1979: "Details and Engineering Analysis of the Illinois Lo-Cal House," Small Homes Council–Building Research Council, University of Illinois, Urbana-Champaign IL 61820.

SHIFRIN, C. 1981: "Propellor Airplanes May Make Comeback," *International Herald Tribune* (Paris), p. 4, 17 September.

SHURCLIFF, W. 1980: *Superinsulated Houses and Double-Envelope Houses*, Brick House Publishing Co., Andover MA 01810.

SHURCLIFF, W. 1980a: *Window Shutters and Shades,* Brick House Publishing Co., Andover MA 01810.
SMITH, V.K. (ed.) 1979: *Scarcity and Growth Reconsidered,* Resources for the Future/ Johns Hopkins University Press, Baltimore, Maryland.
SOCOLOW, R.H. (ed.) 1978: *Saving Energy in the Home,* Ballinger, Cambridge MA 02138.
Soft Energy Notes: bimonthly periodical available by subscription, or as back issues, from IPSEP (International Project for Soft Energy Paths), 124 Spear St., San Francisco CA 94105. The first seven issues (March 1978–July 1979) were reprinted by USDOE, DOE/PE-0016/1, October 1979, and are available from the Technical Information Center, Oak Ridge TN.
Soft Energy Notes 1978–80: end-use data from Hong Kong, Bangladesh, Sudan, Mexico City, and an Indian village, respectively in **1**:21–22, June 1978; **1**:53–58, August 1978; **1**:72–74, October 1978; **2**:20–22, May 1979; and **2**:53–54, July 1979.
Soft Energy Notes 1978a: "Problems and Prospects of Indian Biogas," **1**:10–12, March; "Biogas Powered Piggery in the Philippines," :45–46, August; "Biogas Technology Reviewed," :81–83, October.
Soft Energy Notes 1979: "Fuel Efficient Stoves for Rural Households," **2**:89–90, December.
Soft Energy Notes 1979a: "Soft Energy in China," **2**:80–81, December.
Soft Energy Notes 1980: auto efficiency feature, **3**(4):2–5, August/September.
Soft Energy Notes 1980a: "Heat Engines: An Expanding Technology," **3**(2):24–27, April.
Soft Energy Notes 1981: "Ethanol's Balance Sheet," **3**(6):18–19, December 1980/January 1981.
Soft Energy Notes 1981a: "Affluence from Effluents," **4**(1):10–12, February/March.
SØRENSEN, B. 1979: "Global Energy Policy and Development Strategy," Niels Bohr Institute, Copenhagen; now at Roskilde University Center, Energy Group, Bld. 17-2, IMFUFA, PO Box 260, 4000 Roskilde, Denmark.
SØRENSEN, B. 1979a: *Renewable Energy,* Academic Press, New York.
SØRENSEN, B. 1980: "An American Energy Future," report to SERI from Niels Bohr Institute, Copenhagen, August; see new address above.
SØRENSEN, B. 1981: summary in *Soft Energy Notes* **4**(2):53–55, April/May of "Nordisk Energisamarbete—möglighter och begränsinger i ett långsiktikt perspektiv."
Southern California Edison Co. 1980: Press announcement by Chairman William Gould, 17 October, and 16 October memorandum to employees, reprinted in *Soft Energy Notes* **3**(6):29–30, December 1980/January 1981.
Statistisches Jahrbuch der Eisen- und Stahl-Industrie 1975, Industrieverband Eisen- und Stahl-Industrie, Essen.
STEEN, P. & WIMAN, B. 1977: "Miljövårdens Energibehov," Secretariat for Future Studies, Stockholm.
STEEN, P., JOHANSSON, T.B., FREDRIKSSON,R. & BOGREN, E. 1981: *Energi—*

til vad och hur mycket?, LiberFörlag, Stockholm.
STEERS, L.L. & SALTZMAN, E.J. 1977: "Reduced Truck Fuel Consumption Through Aerodynamic Design," *Journal of Energy* **1**:312.
ST. GEOURS, J. 1979: "In Favour of an Energy-Efficient Society," report to the Commission of the European Communities, Brussels.
STOBAUGH, R. & YERGIN, D. (eds.) 1979: *Energy Future: Report of the Energy Project at the Harvard Business School,* Random House, New York.
STUDIEK, H. et al. 1976: "Substitution und Rückgewinnung von NE-Metallen in der BRD," Institut zur Enforschung technologischer Entwicklungslinien, Hamburg, Januar.
SUDING, P.H. 1980: "Detaillierung des Energieverbrauchs in der Bundesrepublik Deutschland, Teil 1: Haushalte und Kleinverbraucher," Energiewirtschaftliches Institut, Köln, Dezember.
SWEDISH ENERGY R&D COMMISSION 1980: *Energi i Utveckling—Program för Forskning Utveckling och Demonstration inom Energiområdet 1981/82–1983/84,* EFUD 81, SOU 1980:35, Stockholm. (The proposed budget of SKr805M is 36% efficiency improvements, 40% renewables, 3% district heating, 6% fission [safety/wastes], 8% other nonrenewables, and 7% overheads.)
TATOM, J.W., COLCORD, A.R., KNIGHT, J.A. & ELSTON, L.W. 1976: "Clean Fuels from Agricultural and Forestry Wastes," April report to US Environmental Protection Agency, Office of R&D, Washington DC 20460, contract #68-02-1485, from Engineering Experiment Station, Georgia Institute of Technology, Atlanta GA 30332.
TAYLOR, T.B. 1979: *Prospects for Worldwide Use of Solar Energy,* unpublished report to Rockefeller Foundation, New York, from Dept. of Aerospace & Mechanical Sciences, Princeton University; Dr. Taylor is now at 10325 Bethesda Church Rd., Damascus MD 20750.
TAYLOR, V. 1979: *Energy: The Easy Path,* Union of Concerned Scientists, 1384 Massachusetts Avenue, Cambridge MA 02238.
THERMA-WETTBEWERB 1977: "Realisierung des Wettbewerbs THERMA," Bundesministerium für Raumordnung, Bauwesen, und Städtebau, Bonn.
THERMO-ELECTRON CORPORATION 1977: "A National Policy for Industrial Energy Conservation," Rep. #800-135-77, April, Waltham, Massachusetts.
TRW 1979: "Appendix—Data Base on Automobile Energy Conservation Technology," Energy Systems Planning Division, TRW Inc., McLean VA, 25 September.
TSUCHIYA, H. 1980: "A Soft Path Plan for Japan," Research Institute for Systems Technology, 403 Taiyo Bldg., 3-3-3 Misaki-cho, Chiyoda-ku, 101 Tokyo; summarized in *Soft Energy Notes* **3**(3):21–23, June/July, and expanded in *Energy Cultivating Civilization* (in Japanese), Toyo Keizai Shimposha, Tokyo, 1980.
TSUCHIYA, H. 1981: "Efficient Energy Use and Renewable Energy in Japan," paper to Second International Conference on Soft Energy Paths, Rome, January, excerpted in *Soft Energy Notes* **4**(1):8–9, February/March.

TUROWSKI, R. 1977: "Entlastung der Rohstoff- und Primärenergiebilanz der Bundesrepublik Deutschland durch Recycling von Hausmüll," KFA Jülich GmbH, Ber. 1453, August.
UMWELTBUNDESAMT 1979: "Materialien zum Abfallwirtschaftsprogramm 1975 der Bundesregierung," Berlin.
US DEPARTMENT OF ENERGY 1980: $CO2$ 1(3), Status Report, Carbon Dioxide Effects Research & Assessment Program, July, USDOE, Washington DC.
US DEPARTMENT OF ENERGY 1980a: *Workshop on Environmental and Societal Consequences of a Possible CO2-Induced Climate Change*, CONF-7904143, Carbon Dioxide Effects Research & Assessment Program Report #009, October, USDOE, Washington DC 20585.
US DEPARTMENT OF ENERGY 1980b: *Proceedings of the Carbon Dioxide and Climate Research Program Conference*, CONF-8004110, Carbon Dioxide Effects Research & Assessment Program Report #011, December, USDOE, Washington DC 20585.
US DEPARTMENT OF ENERGY 1980c: *Environmental and Societal Consequences of a Possible CO_2-Induced Climate Change: A Research Agenda*, DOE/EV/10019-01/02, 2 vols., Carbon Dioxide Effects Research & Assessment Program Report #013, December, USDOE, Washington DC 20585.
US DEPARTMENT OF ENERGY 1980d: *Low Energy Futures for the United States*, DOE/PE-0020, June, Deputy Assistant Secretary for Conservation & Renewable Resources/Assistant Secretary for Policy & Evaluation, USDOE, Washington DC 20585.
US DEPARTMENT OF ENERGY 1981: *Superinsulated Houses: Least Cost Methods for Space Heating in New Home Construction*, review draft, Office of Policy, Planning, and Analysis, USDOE, Washington DC 20585.
UNIVERSITY OF SASKATCHEWAN 1979: *Low Energy Passive Solar Housing Handbook*, Division of Extension & Community Relations, U. Sask., Saskatoon, Canada.
VON HIPPEL, F. 1980: "Forty Miles a Gallon by 1995 at the Very Least!", PU/CEES-104, Center for Energy & Environmental Studies, Princeton University, Princeton NJ 08540.
WAES (Workshop on Alternative Energy Strategies) 1977: *Energy: Global Prospects 1985-2000*, McGraw-Hill, New York.
WANG, D.I.C. 1981: paper to Toronto meeting, American Association for the Advancement of Science, reported in *Technology Review* 83(5):88, April.
WEINBERG, A.M. et al. 1979: *Economic and Environmental Impacts of a U.S. Nuclear Moratorium 1985-2010*, Institute for Energy Analysis (Oak Ridge TN), MIT Press, Cambridge MA 02139.
WILKINSON, K.G. 1977: "The Role of Advancing Technology in the Future of Air Transport," paper to Royal Society for the Encouragement of Arts, London, 03-02-1977, from Engineering Director, British Airways.
WILLIAMS, R.H. 1978: "Industrial Cogeneration," *Annual Review of Energy* 3: 313-356, Annual Reviews Inc., Palo Alto, CA 94306.
WORCESTER [MASSACHUSETTS] POLYTECHNIC INSTITUTE 1978: *Proceedings of*

the First New England Site-Built Solar Collector Conference, Department of Mechanical Engineering, May.

WORLD ENERGY CONFERENCE 1977: *World Energy Resources 1985–2020,* Executive Summary and full report to the Conservation Commission of the World Energy Conference, IPC Science & Technology Press, UK.

ZIMEN K. et al. 1977: "Source function of CO_2 and future CO_2 burden in the atmosphere," *Zeitschrift für Naturforschung* **32a:** 1544–1554.

About the Authors

Amory B. Lovins, M.A., DD.Sc.*h.c*., a former Oxford don who recently returned to the US after 14 years in England, is a physicist working on energy and resource policy (including, since 1970, energy/climate interactions) in more than 15 countries. He has published nine books and many technical papers, has been a consultant to a wide range of governments and international and private organizations, and has been Regents' Lecturer in resource policy and in economics at the University of California. He served in 1980–81 on the US Department of Energy's senior advisory board and, jointly with his wife and colleague Hunter, with whom he works as a team, is a policy advisor to Friends of the Earth, a US nonprofit conservation group.

L. Hunter Lovins, B.A., J.D., is a lawyer, political scientist, sociologist, and forester. She is a member of the California Bar and a partner of Hirshtick & Sheldon in Los Angeles; has lectured and consulted widely on energy policy, environmental education, and community organizing; co-founded and for seven years was Associate Director of the California Conservation Project ("Tree People"), which is reforesting the Los Angeles area; and has co-authored and edited many recent works with her husband, including their Summer 1980 *Foreign Affairs* article on proliferation, its book expansion *Energy/War: Breaking the Nuclear Link*, and a current study, *Energy Policies for Resilience and National Security*, for the U.S. Federal Emergency Management Agency. In 1982 the Lovinses will be Luce Visiting Professors of Environmental Studies at Dartmouth College and will teach at the University of Colorado (Boulder).

Florentin Krause was raised in the Federal Republic of Germany, but took his Ph.D. in physical chemistry at the University of California at Berkeley in 1976. He has worked as an energy analyst for the ÖKO-Institut in Freiburg, a public interest research group, and was principal researcher and author of *Energiewende*, the first detailed study of a low-energy, high-prosperity future for the Federal Republic. He is currently associated with the International

Project for Soft Energy Paths (IPSEP), a nonprofit San Francisco educational foundation providing up-to-date technical data on energy efficiency and appropriate renewable sources to more than 80 countries.

Wilfrid Bach has since 1975 been Director of the Center for Applied Climatology and Environmental Studies and the Department of Geography at the Westfälische Wilhelms-Universität in Münster, FRG. He studied at the University of Marburg, received his doctorate from the University of Sheffield in England, and has taught at several North American universities and at the Federal Polytechnic of Zürich. He is the author of numerous books and technical papers on climatic change and related issues, and has organized and edited the proceedings of several of the leading international meetings on climate, energy, and food.